Advances in Karst Science

Series Editor
James LaMoreaux, Tuscaloosa, AL, USA

This book series covers advances in the field of karst from a variety of perspectives to facilitate knowledge and promote interaction thereby building stepping stones in this process. Methodologies, monitoring, data analysis and interpretation are addressed throughout the wide range of climatic, geological and hydrogeological contexts in which karst occurs. Case studies are presented to provide examples of advancement of the science.

Issues to be addressed include water supply, contamination, and land use management. These issues although occurring on a local basis share many of the same features on the global stage. This book series is a critical resource to the scientific community allowing them to compare problems, results, and solutions. The presented information can be utilized by decision makers in making decisions on development in karst regions. Contributions presented may be used in the classroom and to work with stakeholders, scientists, and engineers to determine practical applications of these advances to common problems worldwide.

The series aims at building a varied library of reference works, textbooks, proceedings, and monographs, by describing the current understanding of selected themes. The books in the series are prepared by leading experts actively working in the relevant field. The book series *Advances in Karst Science* includes single and multi-authored books as well as edited volumes. The Series Editor, Dr. James W. LaMoreaux, is currently accepting proposals and a proposal document can be obtained from the Publisher.

More information about this series at http://www.springer.com/series/15147

Eric Gilli

Big Karst Chambers

Examples, Genesis, Stability

Springer

Eric Gilli
Department of Geography
Université Paris 8
Saint-Denis, France

ISSN 2511-2066 ISSN 2511-2082 (electronic)
Advances in Karst Science
ISBN 978-3-030-58731-4 ISBN 978-3-030-58732-1 (eBook)
https://doi.org/10.1007/978-3-030-58732-1

This Springer imprint is published by the registered company Springer Nature Switzerland AG
The registered company address is: Gewerbestrasse 11, 6330 Cham, Switzerland

Preface

This book is mainly a translation of a doctoral thesis work in France:

GILLI E., 1984—Recherches sur le creusement et la stabilité des grands volumes karstiques souterrains. Applications au creusement des cavités artificielles de grande taille, Doctoral thesis, Provence University, Marseille, 287 p., 140 fig.).

Therefore, data were actualized, and the thesis work was complemented with data on Sarawak Chamber, the World's largest underground chamber, and on a few other examples out of France. I have also added considerations on the artificial underground large volumes.

The original study was supported by EDF (Electricité de France), the French national electricity company for future projects of digging large underground artificial chambers in view to welcome nuclear power plants. It is easy to create large volumes by digging long tunnels, deep pits, or by hollowing chambers with many pillars but creating a large-span chamber is an important problem for civil engineers.

In the 1980s, Robert Therond, the head of the EDF geological service, was also a caver. He had noticed that some natural caves were very large (250 m in diameter for La Verna Chamber), while human achievements in limestone were only a 37 m span! He decided to finance this work to see if specific geological environments could explain such a difference.

Of course, the idea was not to install a nuclear plant in a cave but to have a naturalistic approach by observing the natural contexts that made possible the existence of large spans.

I had a post-graduate degree in geology and was looking for a work placement or a research job concerning karst systems. One day I knocked at the door of EDF geological service in Paris and met Robert Thérond who told me he could entrust me with studying mountain lakes that were used to produce electricity in connection with nuclear power plants. Indeed, it is not possible to stop a nuclear station; thus, the electricity is used to fill in lakes and later produce hydroelectricity. However, after a few minutes of discussion he also evoked a topic on large underground volumes that he had in mind for several years:

- *It is for us an interesting point of research because I am afraid that one day we will be asked to build an underground nuclear power plant. We do not know how to create an underground volume large enough for doing that. While visiting several caves I have seen that natural examples were large enough and I would like to study them.*

He explained to me that the project was not to install a nuclear plant inside natural caves but to check if a peculiar geological environment or specific forms could explain their large size. Engineers could then use this knowledge for digging artificial cavities. He told me:

- *I wish for a naturalistic approach; and I am looking for someone who is a geologist, geographer, speleologist and specialist in rock mechanics.*

It was a perfect subject for me; however, unfortunately, I did not have the whole requested qualifications. I tried to say:

- *Of course, such a person is rare, but please allow me to say that I am a geologist, geographer and speleologist! I have three of the requested competences. Are you sure you can find someone with four ?*
- *Well.... Let me think about it. I will give you an answer in the afternoon.*

I was merely sure that he would not, but he did!

A few days later, I was recruited with a research contract on the large underground volumes with the possibility of completing a doctoral thesis.

A first surprise of the study was to discover that the topic had not previously been studied. Indeed, most cavers interested in exploring caves wished it to be the longest or the deepest one; thus, when they discovered an underground chamber they usually only measured its length and just said "it is big"! An important part of the work had to be devoted to the survey of the chambers.

A second surprise was to discover the meanders of administration. Everything was forbidden or impossible: a student must work in an office in Paris; a student cannot travel with planes; a student cannot go caving; and a student cannot buy this kind of equipment. We were far from the reality of the fieldwork in caves. Nevertheless, my director and I could navigate in these meanders to organize the job in a proper way.

For doing that job, I spent 2 years traveling all around France with caver friends that were happy to join this new adventure. We were young, with no money and this thesis was an incredible opportunity to discover new places and new caves, sharing comfortable hotel rooms instead of cold and wet tents.

On May 25, 1984, I presented the results of that study and obtained my Ph.D. degree as doctor in geology. During the work, I had heard about the Sarawak Chamber in Malaysia that had been discovered by a British team of cavers in Mulu, a small mountain in Borneo Island. Of course, the EDF administration had not allowed me to go there, and I was curious to see if my conclusions did apply to that huge chamber. I tried to find financial support. I was only able to make a recognizing trip in 1986 with a grant from National Geographic Magazine, but the heads did not accept to finance the project itself. In 1993, I presented it to the Rolex Awards for Enterprise, and I received an honorific nomination; however, despite this honor, I could not get enough money to realize the study.

Finally, with three friends of mine, we decided to go for an auto-financed project. Collecting money in our pockets and doing small jobs made it possible to finance four plane tickets and the necessary amount to reach Mulu, but no more. The initial objective had to be reduced. Thus, instead of making an accurate study, we decided to focus on a photography project. For doing that, we had trained in France in large chambers or by night in limestone quarries. After a few months, we were ready to explore the darkness of Sarawak Chamber. We did our project and could see that the conclusions were confirmed.

Saint-Denis, France Eric Gilli

Acknowledgements I would like to thank the Geological Service of Electricité de France (French Electricity Company) for providing the financial support for this thesis, and especially Robert Thérond who initiated the subject and who led this research. My thanks also go to Claude Rousset my director at Marseille University and to the owners of the show caves described in this work who welcomed us.

The Sarawak project (Chap. 5) received a grant from National Geographic Magazine in 1986 and a special mention from the Rolex Awards for Enterprise in 1987. I wish to express here my deep gratitude.

I will not forget, of course, all my caver friends who have agreed to accompany me during the field trips and without which this work could not be done. Their list is at the end of Chap. 2.

Thesis Abstract

As part of a research contract with Electricité de France, 28 karst cavities of more than 50 meters (m) reach were studied. For each, a precise map, a geological survey and observations on their stability were made using light field methods. The study shows that 450 m spans are possible. Three modes of natural digging are considered:

- by dissolution;
- by dissolution and collapse;
- by scouring-racking and collapse.

The second part is devoted to artificial cavities and to the definition of various geological environments likely to make possible the digging of large artificial cavities.

Résumé de la thèse: Dans le cadre d'un contrat de recherche avec Electricité de France, 28 cavités karstiques de plus de 50 mètres de portée ont été étudiées. Pour chacune, un plan précis, un lever géologique et des observations pour la stabilité ont été effectués avec des méthodes légères. L'étude montre que des portées de 450 mètres existent. Trois modes de creusement sont envisagés:

- par dissolution;
- par dissolution et effondremen;
- par affoui llement-soutirage et effondrement.

La deuxième partie est consacrée aux cavités artificielles et à la définition de divers environnements géologiques susceptibles de permettre des réalisations de grande taille.

Contents

1 Presentation

This chapter presents the objective of thesis work on large underground karst volumes: giant galleries; rooms, chambers, halls, collapse shafts and sinkholes; and entrance porches. The dimensions of several examples are larger than artificial cavities dug in limestone. The aim of the fieldwork is to collect data that could be used by engineers to hollow out artificial caverns large enough to shelter nuclear power plants. Due to poor bibliographic data, an important fieldwork was necessary. Most of the caves had to be surveyed. Specific equipment and methods were developed to make it possible.

1.1 Introduction

1.1.1 The Large Underground Volumes

The exploration of the subterranean world sometimes places the observer in the presence of underground volumes of a magnitude beyond the size of the artificial cavities that are realized to date. In the 1950s, during the exploration of the Pierre Saint-Martin Cave network (Pyrenees, France), after several hours of walking in a gallery crossed by a river, the speleologists arrived in a place that was so vast that they first believed they had reached an exit of the cave, at nighttime and that clouds prevented the stars from being seen. Meanwhile, their watches indicated a daylight time! They had to quickly admit that they were in a gigantic room which was named Salle de la Verna (La Verna Chamber) (Attout 1954). Its dimensions, 280 m long, 230 m wide and about 180 m high, made it the largest underground hall known in the world. This world record fell in 1958 with the discovery of the Torca del Carlista in Spain (497 m long, 285 m wide and 197 m high) (Courbon 1975) (Fig. 1.1) and then the Sarawak Chamber in Malaysia by the BCRA in 1980 (650 m long, 409 m wide and 98 m high) (Waltham and Brook 1980) (Fig. 1.2). The latter being the world's largest one even if some people consider that the Miao Chamber in China (Vergano 2014) or the recently discovered Hang Son Doong in Vietnam could be larger (Bisharat 2018). However, it depends on the way the chamber is measured (volume, surface or width).

If one compares the size of these natural examples with the maximum dimensions of artificial cavities that are dug in similar terrains, that is to say, those of the Sautet underground plant (France, 33 m wide) (Fig. 1.2), he finds that natural chambers are ten times in range and several hundred times in volume larger than artificial caverns. Moreover, these cavities, although without any pillar present, in the majority of cases, show a surprising stability.

1.1.2 Objective of Studying the Underground Large Volumes

The Geological and Geotechnical Division of Electricité de France (French Electricity Company) had been confronted for many years with the problems of digging underground power plants and large galleries. It was therefore interesting to undertake, as part of a doctoral thesis of geology, the study of the great French natural cavities. The first step in such a study was to carry out a "naturalistic" analysis (survey, morphology, geology, genesis), which could serve as a basis for later geotechnical studies.

The present work represents the results of this first step. In this context, a list of large underground cavities was established. For the French territory alone, 36 caves are wider than 50 m. The word "wider" means that the ceiling span of the cave passages (galleries or chambers) is more than 50 m wide; indeed, caves can be several kilometers (km) long and usually speaking of a big or a large cave is a very long one. Most of the cavers are more interested in length and depth of cave networks than in the width of passages. The study focused on 21 of these large-span cavities. A few cases of slightly smaller dimensions have, however, been studied because of their interest. Three large

E. Gilli, *Big Karst Chambers*, Advances in Karst Science,
https://doi.org/10.1007/978-3-030-58732-1_1

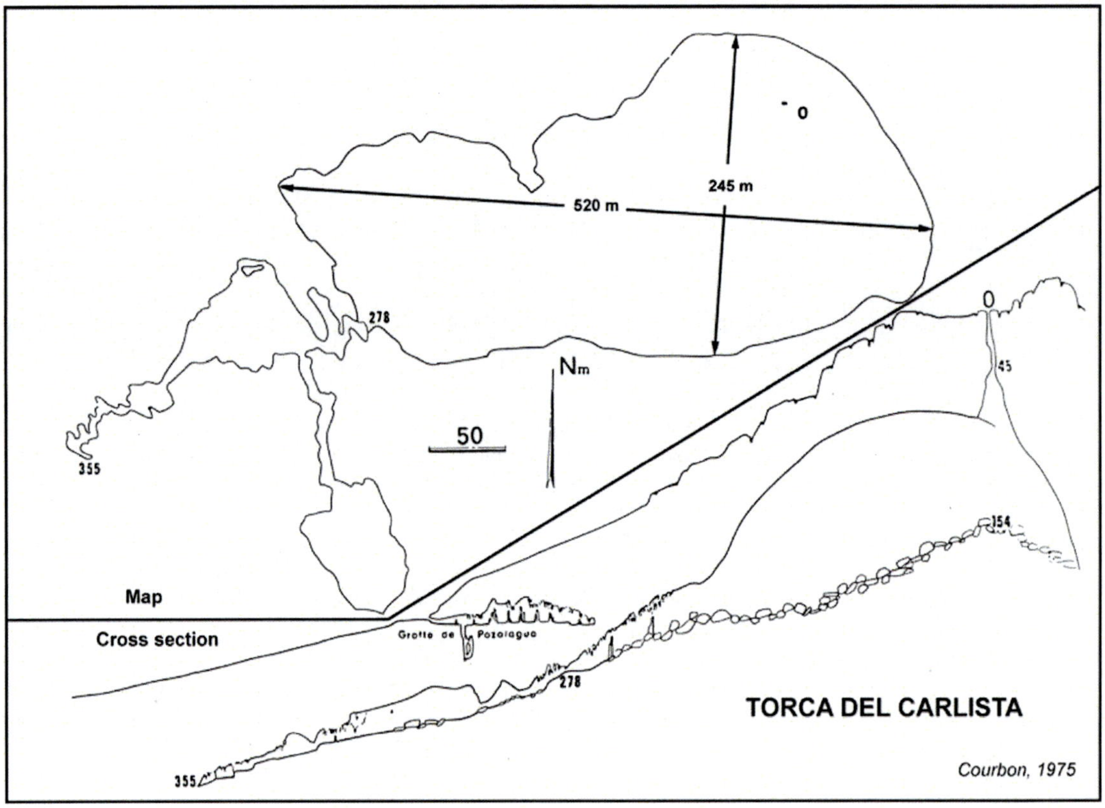

Fig. 1.1 Torca del Carlista, Europe's largest chamber

caves located in Turkey and the Sarawak Chamber in Malaysia were also presented (Fig. 1.3).

1.1.3 New Data Since the 1984 Thesis Work

Since this thesis work, as far as I know, few works have been realized in the same topic. In the 1990s, I tried to find financial support to study the Sarawak Chamber. Unfortunately, in spite of a small grant from National Geographic which made it possible to do a recognizing trip in 1985 and a Rolex Awards nomination in 1993, I was unable to organize an important project. However, I could convince three friends of mine to join me for a short trip to Borneo in 1993. The aim was to take a picture of the chamber and to make geological observations. I present here the data collected during that short field trip.

Since that time, new chambers were discovered in other countries like China or Vietnam, and large underground artificial caves exist now that are wider than the Sautet underground plant, for instance, the Tytyri underground quarry in Finland (10,000 m^2 without pillars) (see Table 1.1 and Chap. 6).

1.2 Definition of Large Underground Karst Volumes

In the present study, the large underground karst volumes are classified into four categories:

- the giant galleries;
- the rooms, chambers or halls;
- the collapse shafts or sinkholes;
- the entrance porches.

Each category corresponds to a well-defined type; however, in reality, many examples are difficult to classify. Giant galleries are elongated karst conduits, large in size, predominantly horizontal. They result from the activity of an underground river. In this case, the entire network offers large volumes. After digging, speleothems, collapses or karst may narrow some passages and give to the galleries an appearance of successive rooms. Rooms, chambers or halls are sudden enlargements of a gallery which is generally provoked by the collapse of the roof.

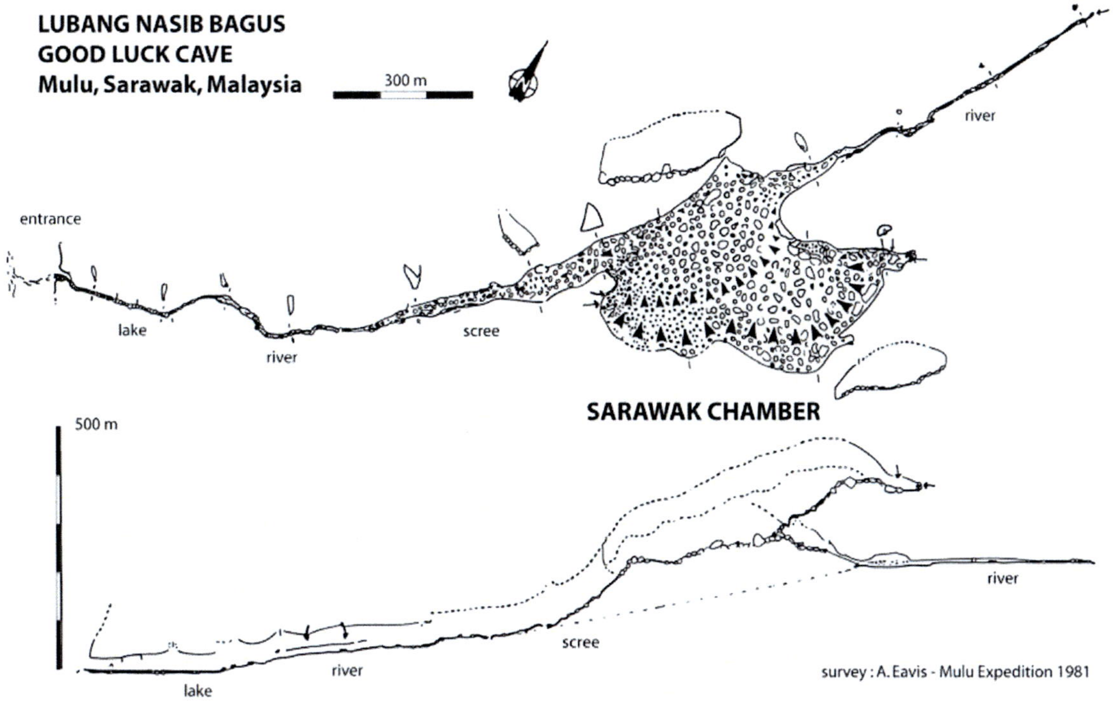

Fig. 1.2 Sarawak chamber, corld's largest chamber

The gallery resumes its initial dimensions downstream of the room. Rooms are morphological accidents within a network. They can also result from intersections between several galleries (Renault 1957). Collapse shafts exist when there is a connection between a collapsed room or a giant gallery and the surface. These are, therefore, special cases of the two preceding classes. Entrance porches are galleries that suddenly widen when reaching the edge of the massif. They are usually at the foot of a cliff and can be considered as half-rooms.

1.3 Method for Studying the Large Chambers

1.3.1 Introduction

Each cavity has been the subject of a bibliographic search in French inventories (Aellen and Strinati 1975; Chabert 1981; Courbon 1980; Lismonde 1978; Martel 1921; Minvielle 1977). In particular, the journals of many French speleological clubs were examined. For most cavities so identified, the published maps were not accurate enough which is very normal given the motivations and means of their authors. It was, therefore, necessary to make a precise survey of the selected caves and to focus on the topography of the ceiling which was generally neglected.

The topographic map was complemented by a geological survey and by the search of signs concerning the stability of the walls and roofs. The chambers were also placed in their geomorphological context on a 1/10000 map.

1.3.2 Means and Techniques

Note: Most surveys were performed by two persons. For this reason and also considering the problems of environment (access, darkness, etc.), it has been necessary to use a simple, light and solid equipment. In the 1980s, laser range meters were too expensive and cave software were not available. Thus, specific equipment and methods had to be developed that are described below. Of course, nowadays, modern equipment and cave mapping softwares improve the quality of the surveys.

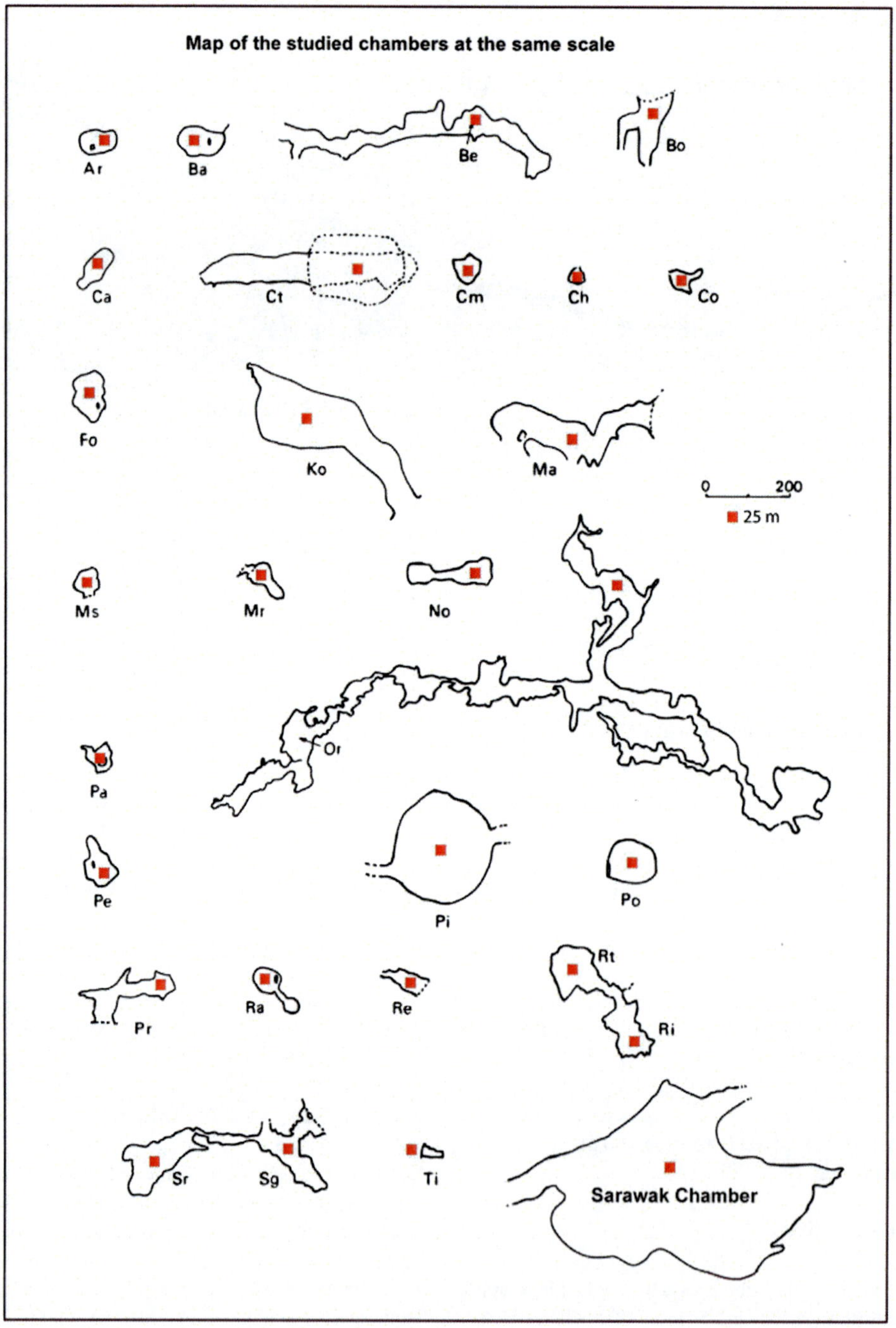

Fig. 1.3 Sketches of the studied caves. The small red square is 20 m × 20 m

Table 1.1 World's largest chambers

Name	Cave	Country	Width	Length	Hight	Surface	Volume
Sarawak	Lubang Nasib Bagus	Malaysia (Sarawak)	**409**	650	98	162,700	9,800,000
Titan chamber	Xiniu	China	**341**	367	80	54,750	2,500,000
Miaos	Gebihe	China	**336**	824	177	140,540	10,800,000
Gran salon	Torca del Carlista	Spain	**287**	497	197	87,090	2,100 000
Main chamber	Ghar-e-Dosar	Iran	**265**	385	70	81,500	5,000,000
Majlis al Jinn	Majlis al Jinn	Oman	**245**	347	?	61,000	4,100,000

1.3.3 Equipment

1.3.3.1 Measurement of the Floor

Equipment:

- 1 double decameter
- 1 Suunto compass
- 1 Suunto clinometer.

The last two devices are sighting equipment. They are used with both eyes open: one is looking at the station, the other eye reads the compass scale. The precision is of the order of half a degree. For short shots, a parallax error exists due to the distance between the eyes. However, in use, these devices were sufficiently accurate.

1.3.3.2 Measurement of the Ceiling

Up to a hundred meters high, the used material was a bright colored balloon inflated with helium and connected to a thin wire. It gives a direct value of the height above the topographic station. The line was a polyester one with a knot glued every meter. A color code made it possible to avoid a one-by-one counting of the nodes. The line was installed on a fishing reel. Another possible method was to use a "topofil" a kind of cavers' thread meter. The measurement then takes a lot of time because the balloon has to be attached again after recovery of the entire thread.

The gas was transported in a small helium cylinder (Model B1 of "Air Liquide" company: 33.5 cm high, 8 cm in diameter, 1.5 kg, pressure 200 bar). It was essential to carry the cylinder into the cave because balloons are fragile, and they sometimes explode when arriving in contact with the asperities of the ceiling. A good projector is necessary to see the position of the balloon on the ceiling.

1.3.4 Method

1.3.4.1 Survey of the Floor

The shape and altimetry of the chamber floor were defined by a path around the room and through several crossings. The room was thus shared in several sectors that each represents so many loops (Fig. 1.4). The accuracy of the work was given by the loop closure error. During this study, despite the simplicity of the devices used, it was possible to obtain a precision of less than 1%.

Each station was materialized by a numbered lit candle. Control measurements from characteristic points made it possible to correct eventual errors. It is wise to choose stations that are located a bit high above the soil (stalagmites, blocks). When slopes are important, measurements must be realized from the low point. Indeed, the compass must be used horizontally; and it is easier to see a station above it than below it. It is this type of measurements which generally generates the most important errors.

Note: It is necessary to carefully check the personal equipment (helmet, glasses, pencils, etc.) in order to eliminate metal that provokes a compass deviation.

1.3.4.2 Survey of the Ceiling

Note: In speleology, the word vault is often used in an abusive way instead of ceiling or roof. I sometimes use this word although it gives no indication on the shape of the ceiling.

The survey is made by performing a height measurement at the vertical of some ground stations, especially when crossing the chambers. The "helium balloon" method is fast and straightforward. It presents, however, several inconveniences:

- the thread is extensible (1–2% error),
- the balloons are fragile,

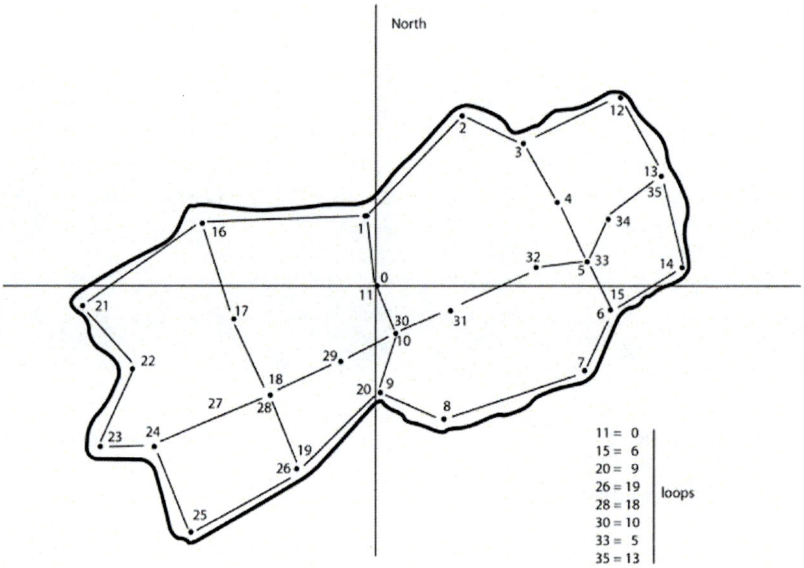

Fig. 1.4 Survey path on the floor of a chamber

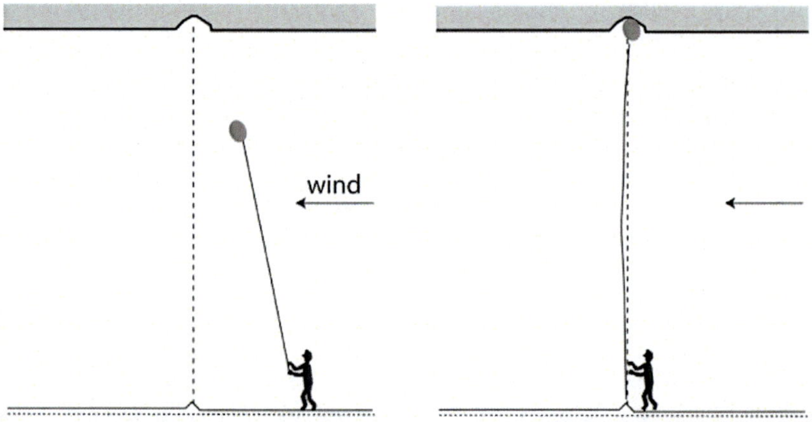

Fig. 1.5 Height measurement in the presence of airflow

- the ascensional force of a single balloon is low; for heights higher than 60–70 m, several balloons must be used to compensate the weight of the wire and the drops of water that fall on the balloons (four for the Verna room),
- air currents, which are frequent, are a nuisance and can prevent the measurement.

In such a case, one can use the following method. A small bell is searched in the ceiling. In its verticality, a station is located on the floor, with the help of the clinometer. Then, the balloon is placed into the bell, and the operator moves to the station marked on the ground while straightening the wire (Fig. 1.5). However, an error remains due to the curvature of the thread under the effect of the wind. Another point to report is that ceilings are not horizontal; thus, the balloon, therefore, tends to drift upward. It is therefore necessary to carry out the measurement from the first contact of the balloon with the ceiling. It is essential to have a powerful flashlight to be able to observe this first contact (Fig. 1.6).

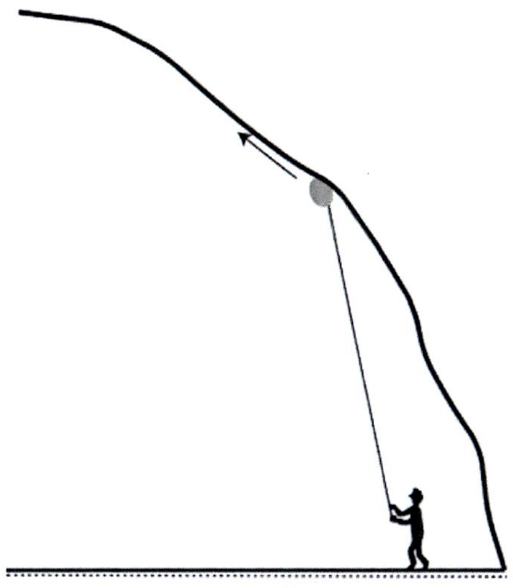

Fig. 1.6 Example of balloon shift on the ceiling

1.3.5 Graphic Restitution

To eliminate the graphic errors, a coordinate's method was used. Each station was spotted in an X-, Y-, Z-axis systems with X = North, Y = East and Z = Altitude. The x- and y-coordinates of each station were cumulated with respect to the first station (No. 0 is usually placed at the entrance of the hall). The dots were then placed on a sheet of graph paper (Fig. 1.7). This method eliminates errors due to devices (protractor, ruler, pencil). It also makes it possible to know directly the rectangle in which the chamber fits by seeking the extreme coordinates.

Nowadays, laser range meters, cave computers and lidar facilitate the work, and 3D cave representation is possible (Jaillet et al. 2019; Zlot and Bosse 2014). La Verna Chamber in France was surveyed with 24 million dots in 2005 by ATM3D society (Chazaly et al. 2010). A new project is in progress.

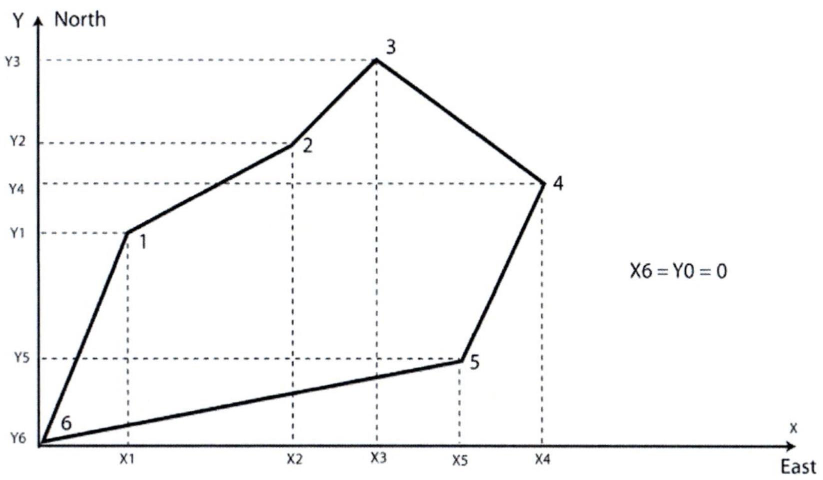

Fig. 1.7 Graphic restitution using coordinates

References

Aellen, V., Strinati, P.: Guide des Grottes d'Europe, p. 320. Delachaux et Niestlé edit, Neuchâtel—Paris (1975)

Attout, J.: Les hommes de la Pierre Saint Martin. Collection Marabout Junior (1954)

Bisharat, A.: Explore the world's biggest cave from your couch. National Geographic News (2018). https://www.nationalgeographic.com/news/2018/05/son-doong-cave-vietnam-virtual-reality-culture Accessed 2 May 2018

Chabert C.: Les grandes cavités françaises. Fédération française de spéléologie edit., pp. 154 (1981)

Chazaly, B., Saillant, M., Varrel, E.: La lasergrammétrie, un nouvel outil pour cartographier les cavités. Actes du colloque AFK—Pierre St. Martin 2007. Karstologia Mémoires **17**, 93–101 (2010)

Courbon, P.: La Torca del Carlista (Espagne), l'une des plus grandes salles du monde. Spelunca **1**, 18–20 (1975)

Courbon, P.: Atlas souterrain de la Provence et des Alpes de lumière, S. C. Sanary, pp. 200 (1980)

Jaillet, S., Delannoy, J.-J., Génuite, K., Hobléa, F., Monney, J.: L'image topographique du karst et des grottes: représentations 2D et technologies 3D, entre réalité et imaginaire. Géomorphologie **25–3**, 191–205 (2019)

Lismonde, B.: Grottes et scialets du Vercors **1** (1978)

Martel, E.A.: Nouveau traité des eaux souterraines. Boin edit, Paris (1921)

Minvielle, P.: Grottes et Canyons. Denseil edit., pp. 234 (1977)

Renault, P.: Effondrements, seismes, et failles vivantes, Ann. Spéléo. CNRS Paris **XII**(1–4), 47–54 (1957)

Vergano, D.: China's supercave takes title as world's most enormous cavern. National Geographic News (2014). https://www.nationalgeographic.com/news/2014/9/140927-largest-cave-china-exploration-science Accessed 27 Sept 2014

Waltham, A.C., Brook, D.B.: Geomorphology in Gulung Mulu Cave. BCRA **7**(3), 123–140 (1980)

Zlot, R., Bosse, M.: Three-dimensional mobile mapping of caves. J. Cave Karst Stud. **76–3**, 191–206 (2014). https://doi.org/10.4311/2012EX0287

Monographs

2

This chapter presents the caves studied during the thesis work. For each chamber, the following data are provided: description, geology and hydrogeology, genesis and stability. For each cave, a map and several cross sections are provided with information on the geological and structural features. Twenty-three caves were surveyed in France and two caves in Turkey. The study includes a few collapsed caves to check the possible causes of unstability.

2.1 Aven Armand (Hures-La-Parade, Lozere, France)

2.1.1 Description

See Figs. 2.1, 2.2, 2.3, 2.4, 2.5, 2.6, 2.7.

Probably known for a long time, the Armand Shaft was explored on September 16, 1897, by Armand, Martel and Viré (Martel 1937). It is now a show cave (André 2017). To reach it from Meyrueis, take the D 986 road toward the northwest. At 10 km on the left, a branch leads to the entrance of the shaft (Coordinates: 44° 13′ 15″ N, 3° 21′ 25″ E).

It is accessible by a funicular, in an artificial gallery, but the natural entrance is a 75-m-deep shaft that gives access to a room which is 85 m long and 57 m wide. It is shaped as a flat ovoid cylinder inclined 20° to the east. The floor of the lower part of the hall is occupied by a remarkable "forest" of stalagmites that causes a greater interest in this show cave (Fig. 2.4). In its upper part, the soil is covered with a scree. The southern part is limited by a large vertical diaclase; it is the same for the western wall. The north wall is disorganized by a double network of joints (see 2.1.4.3). The ceiling has few speleothems and has a regular shape. It is constituted by slabs supported on each other with spans up to a few tens of meters (Fig. 2.1). The final aspect is that of a corbelled vault. At the lowest point of the chamber, an 87-m-deep pitch leads to the end of the cave (Figs. 2.3, 2.4, 2.5, 2.6, 2.7).

E. Gilli, *Big Karst Chambers*, Advances in Karst Science,
https://doi.org/10.1007/978-3-030-58732-1_2

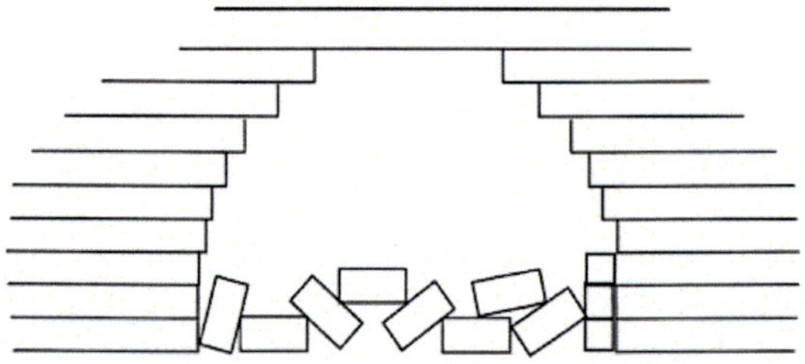

Fig. 2.1 Corbelled vault

Fig. 2.2 Geomorphological map of Armand Shaft

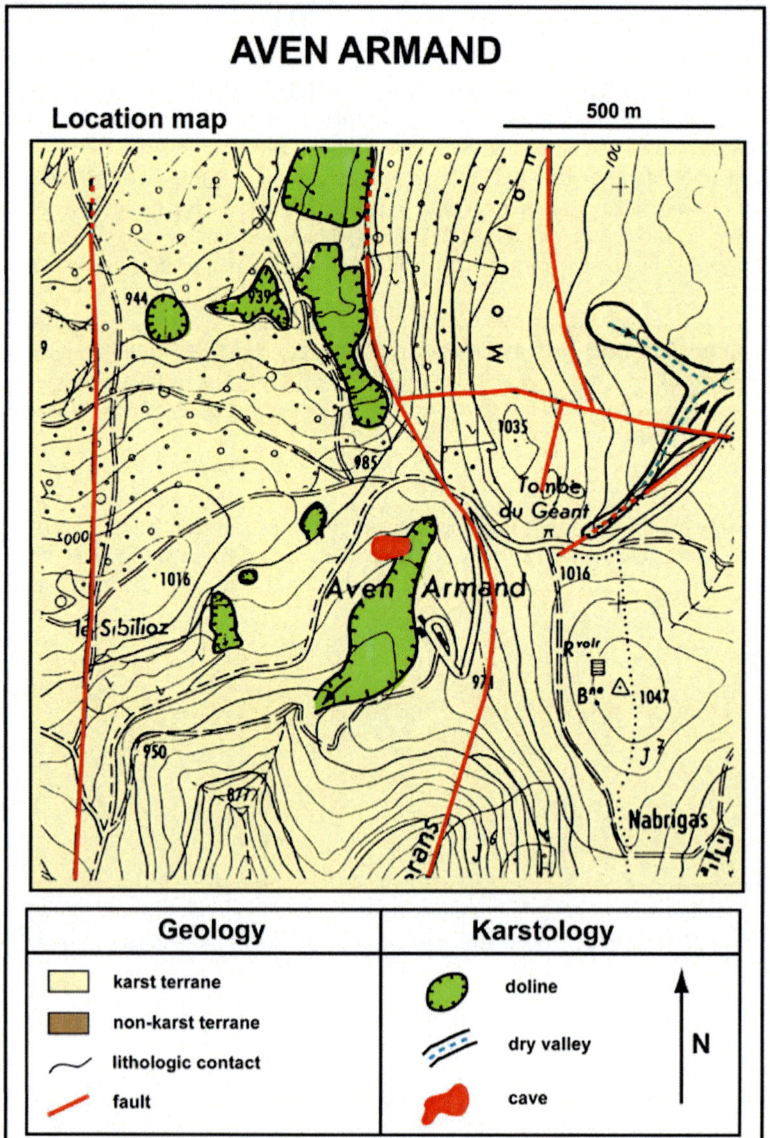

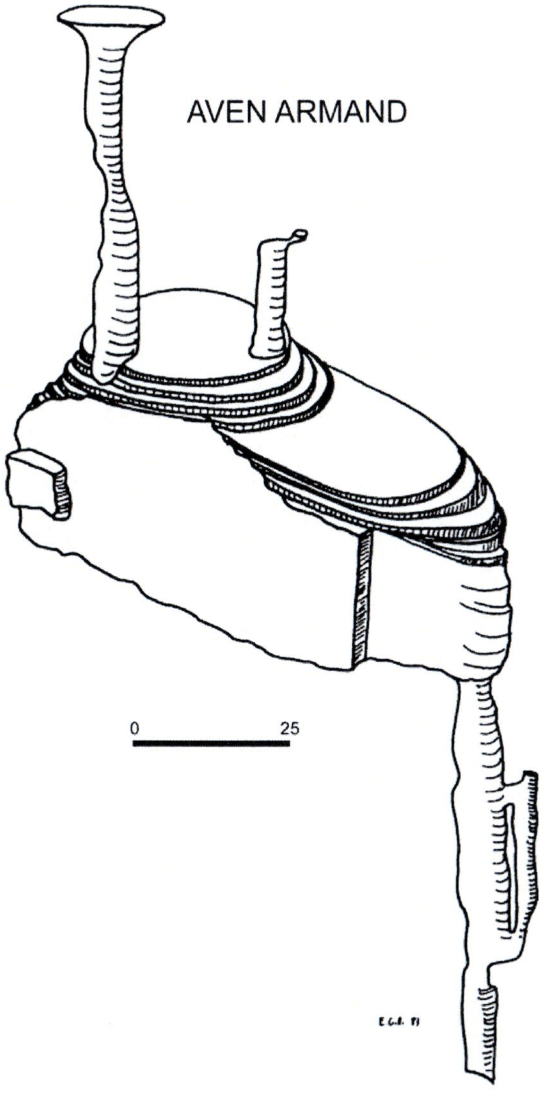

Fig. 2.3 3D sketch of Aven Armand Shaft

2.1.2 Geology and Hydrogeology

The Aven Armand is 970 m above sea level (m.a.s.l.) on the Méjean limestone plateau. The entrance of the chasm is located on a flat area occupied by a half-open doline, cut by the Herans talweg, a tributary of the Jonte River. On its eastern part, the plateau is bordered by an escarpment due to the presence of the Causse Mejean median fault that one can note before arriving at the Speleo Station building (Fig. 2.8). The chamber is located 700 m from the flank of the Herans talweg. The thickness of limestone above the cave is between 45 and 60 m.

The Causse Mejean is a limestone and dolomite plateau with many karst phenomena. It is crossed by two main accidents:

- In the center, the median N-S fault that passes near the shaft.
- In the East, multiple WNW-ESE faults.

The cave develops in several formations:

- The entrance pit is in the Upper Kimmeridgian: very thick beds of white limestone or sublithographic platelets (near the orifice).
- The actual room is located in sublithographic limestone in large beds 1–2 m thick of the Lower Kimmeridgian or the Upper Oxfordian (Rouire and Rousset 1973).
- The terminal pit is likely to intersect the Upper Oxfordian: gray-white sublithographic limestone.

The layers form a very compact assembly with a regular dip, 10° east.

The Aven Armand is located near the median north-south fault of Causse Mejean that dips to the west. The terminal pit is located a few tens of meters from this accident and may even be aligned on it. The chamber is situated in a very fractured area. The most frequent joint orientations are N55° and N165°. Over the entire cave, both networks form a net that determines the shape of the walls.

Fig. 2.4 Stalagmite forest in Aven Armand Shaft (licence Creative Common)

There are three areas of cave fillings (Fig. 2.5).

- The basis of the entrance pit, occupied by a slope of small allochthonous blocks, whose origin is natural or artificial (e.g. stones thrown by the peasants).
- The northwest quarter, consisting of large blocks of various sizes.
- The eastern half of the room is occupied by a "forest" of stalagmites implanted on a giant scree (Fig. 2.4). These stalagmites are often large (from a few meters to more than 20 m) and are old.

The terminal pit is likely hollowed in the blocks in its upper part rather than in the massive rock below. Unlike the floor, the ceiling has few speleothems. The cavity is currently dry. After heavy storms, the water level rises in the final pitch.

2.1.3 Genesis and Stability

The Aven Armand is a collapse room that was formed into a very fractured area in relation to a deep groundwater circulation. The aspect of the filling implies a great stability of the cave. Indeed, the quantity of stalagmites at the back of the room can only be explained by the absence of movement at ground level. The terminal pit therefore no longer plays its role of racking. The southern wall, which is a large diaclase plane, is flat and perfectly stable. The eastern and western walls are fairly regular and stable. The northern wall is disorganized by joints but seems stabilized.

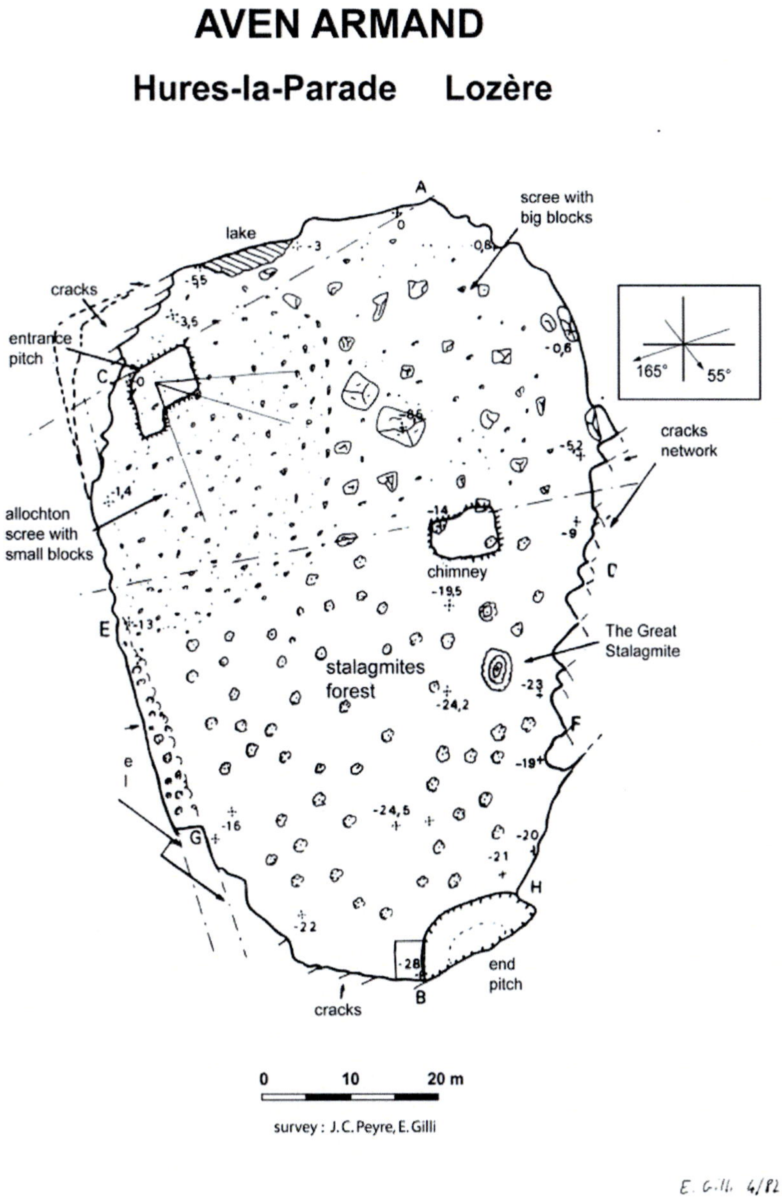

Fig. 2.5 Map of Aven Armand Shaft

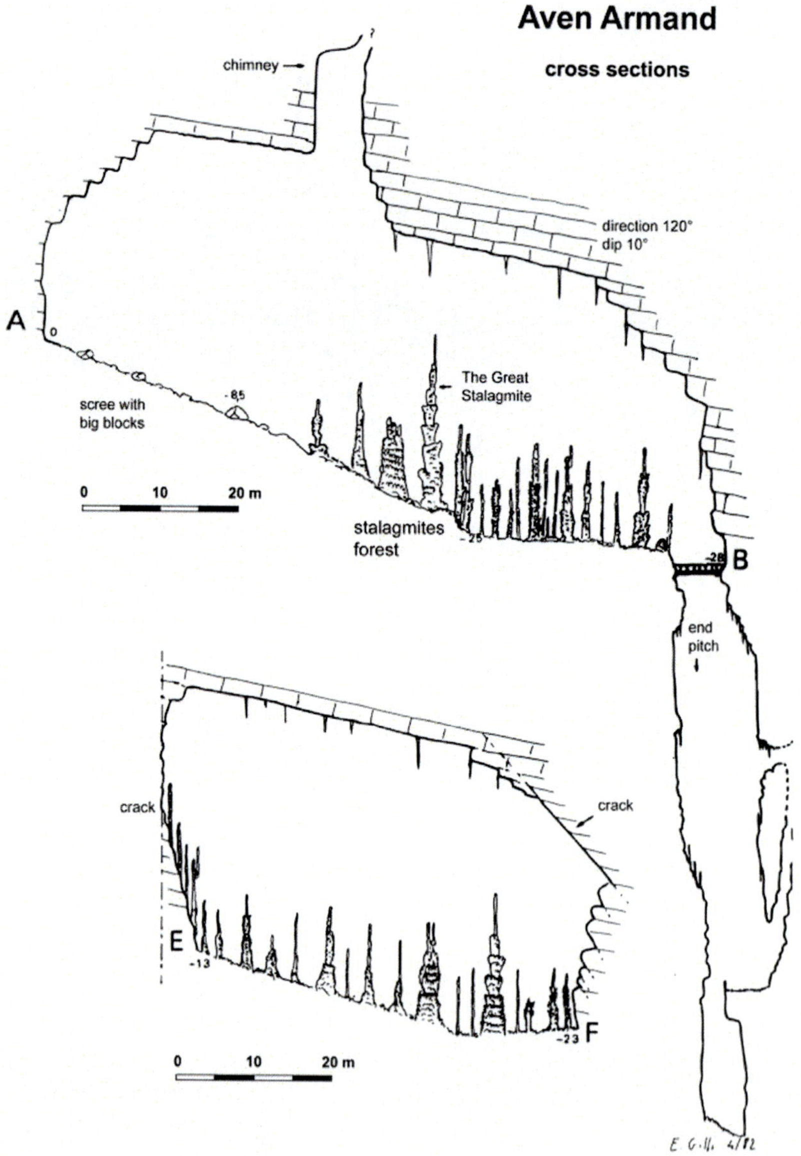

Fig. 2.6 Cross section A–B of Armand Shaft

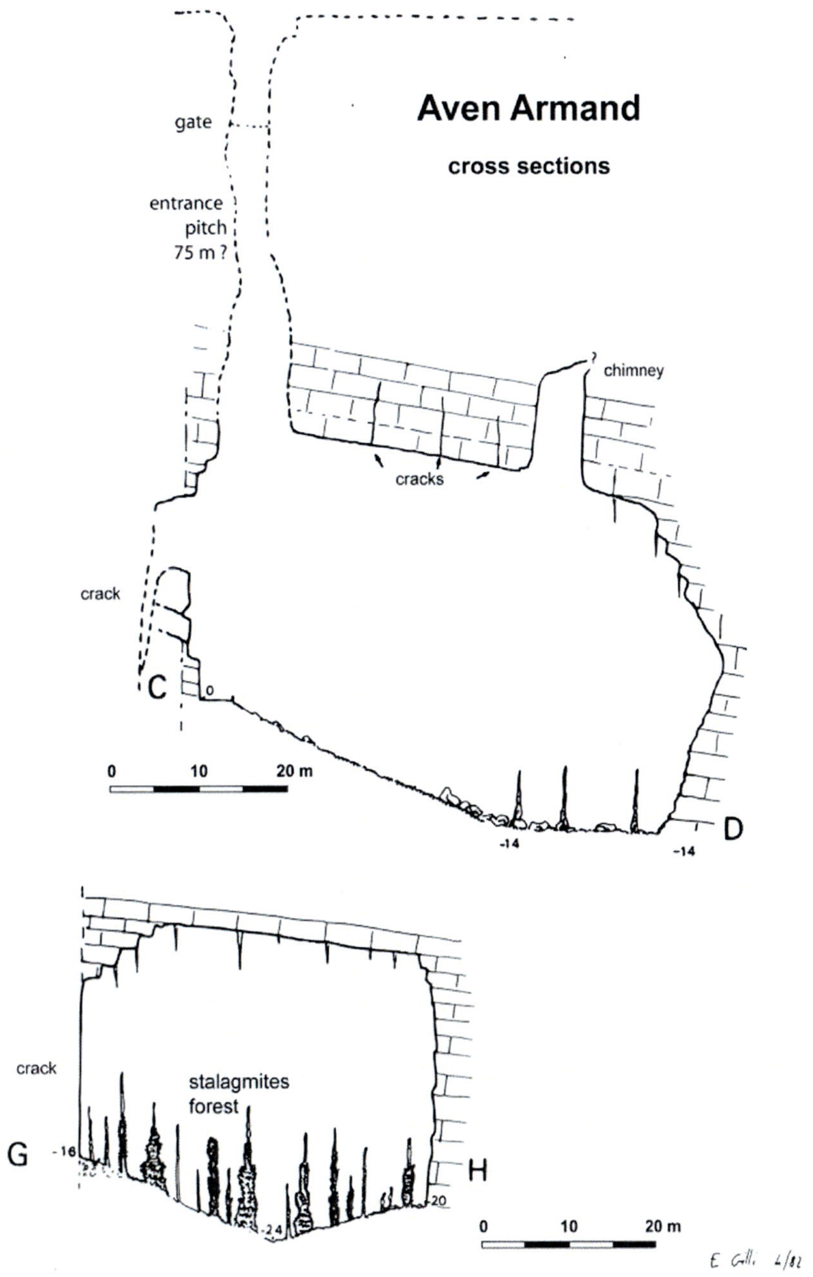

Fig. 2.7 Cross sections C–D and G–H of Armand Shaft

2.2 Baume Sainte Anne Shaft (Sainte Anne, Doubs, France)

2.2.1 Description

See Figs. 2.8, 2.9, 2.10, 2.11, 2.12.

From Nans-Sous-Sainte-Anne (close to the Lison spring and the Sarrazine Cave) drive on the D103 until Crouzet-Migette. From there, take a right to the D229, then 150 m after Sainte Anne, leave the road and climb across a field on the left toward a bunch of trees in the middle of which is the shaft.

The cave is an oval room, 70 m by 120 m in dimension, lengthened along its east-west axis. The entrance is in a collapse sinkhole with an earthy and stony edge that becomes steeper and steeper downward. By climbing down

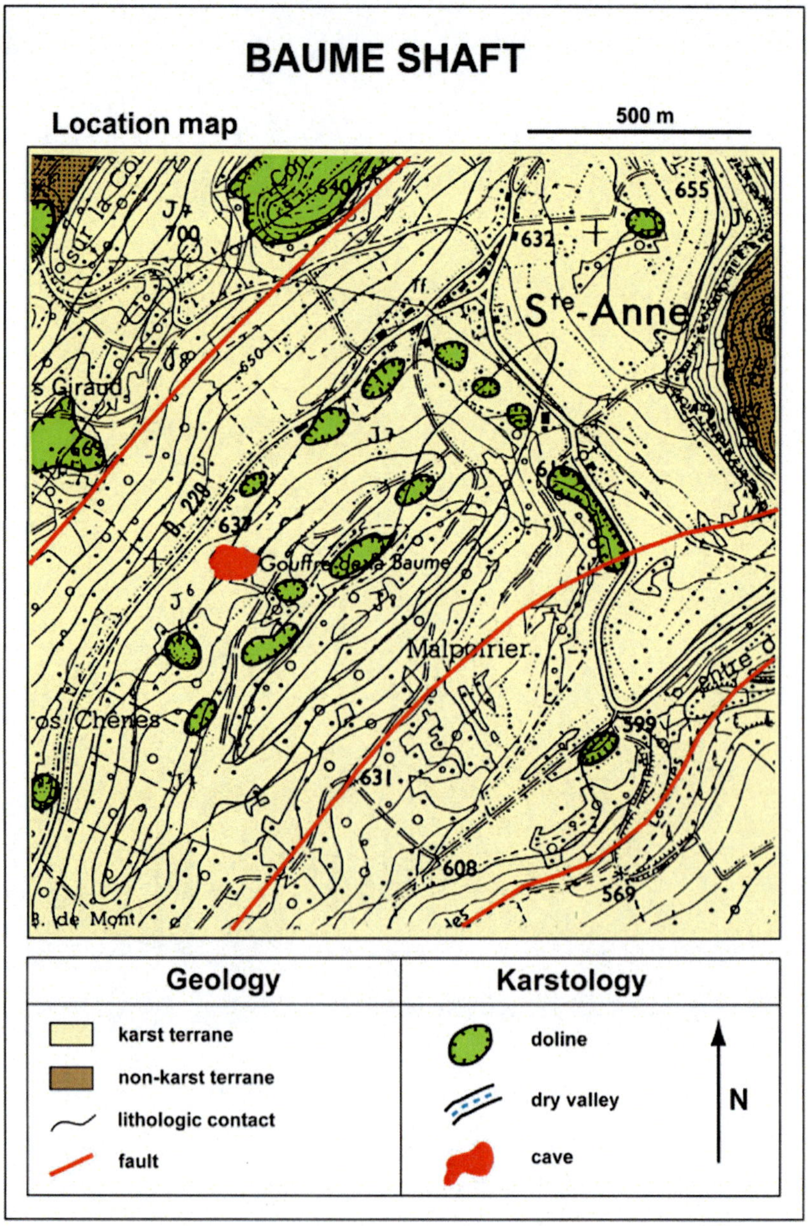

Fig. 2.8 Geomorphological map of Baume Shaft

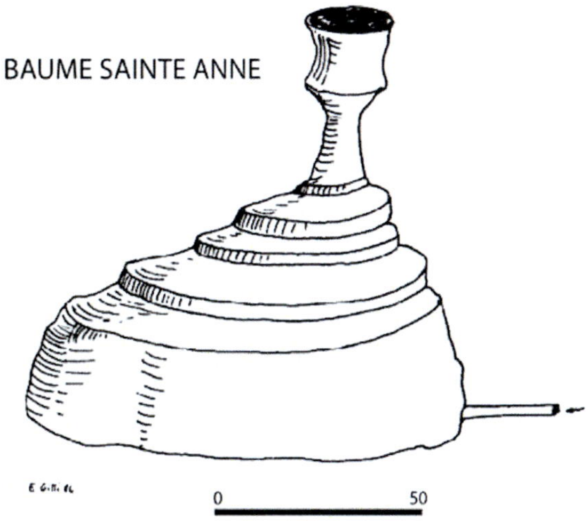

Fig. 2.9 3D sketch of Baume Sainte Anne

Fig. 2.10 Entrance shaft of Baume Sainte Anne (photo Agamemnon)

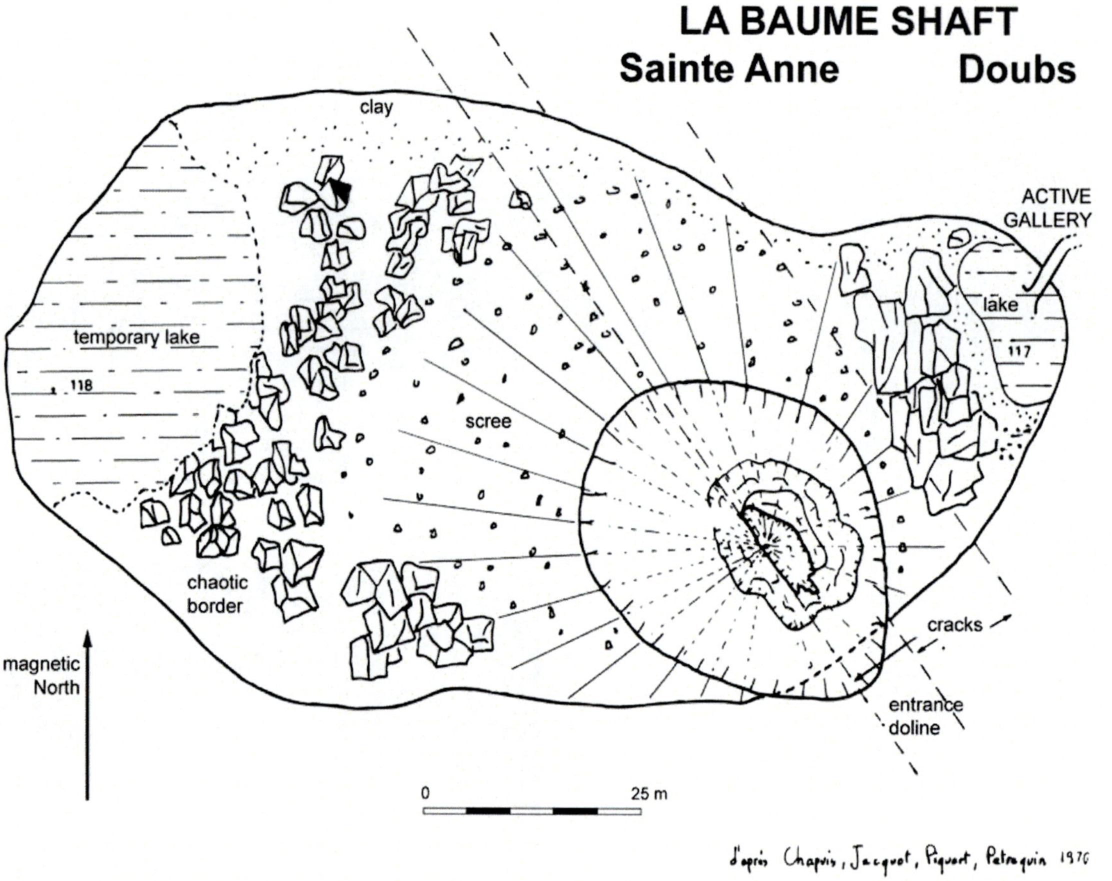

Fig. 2.11 Map of Baume Shaft

on the rope about 20 m, one arrives at a clay path. The shaft then tightens, its walls become vertical, rocky and massive until −40 m where it reaches the ceiling of the chamber. The arrival is at −90 m at the top of a scree cone that occupies most of the cavity. The edges of the chamber are chaotic. Both eastern and western ends are occupied by a lake. The ceiling is a vault; its morphology is influenced by the presence of large limestone beds that give it the appearance of a corbelled vault, as well as by the existence of large tectonic planes, particularly in the eastern part of the room.

2.2.2 Geology and Hydrogeology

The Baume Shaft opens at 650 m altitude on the northern flank of a small hill located 1.5 km west of the Reculée--du-Grouin Gorges (Fig. 2.8). The karst landscape is forested, with numerous sinkholes. The entrance of the shaft is a deep collapse sinkhole enlarged by subsidence. The thickness of limestone above the ceiling is approximately 50 m. The cavity is hollowed out in a horizontal series on the side of a synclinal axis N.E.-S.W. This synclinal is flanked by two parallel vertical faults oriented N30°. The entrance is aligned with a large vertical fracture oriented N140°.

From the top to the base, the lithology is as follows:
Sequanian (Lower Kimmeridgian) (entrance doline)

- 20 m of limestone in shallow banks with more or less marly zones;
- 0.5 m of green marl;
- 50 m of very compact white limestone.

Rauracian (Upper Oxfordian)

- 50 m of sandy limestone with organic debris;
- marls.

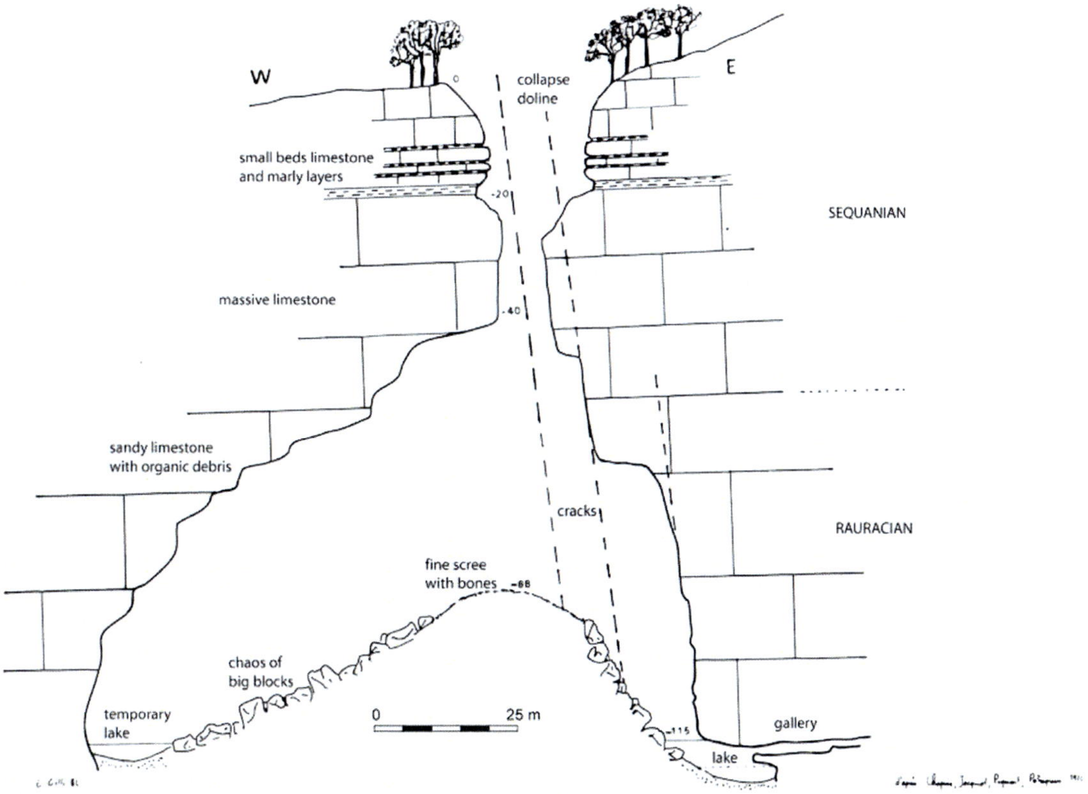

Fig. 2.12 Cross section of Baume Shaft

The presence of the level of green marls, combined with many joints, has probably caused the collapse of the ceiling. This marl's level is emphasized by the cornice that extends 20 m below the surface.

The stratification is horizontal. The large vertical N140° joints, which caused the opening of the entrance pit, belong to a network of fractures that is also present in the eastern part of the room, at the ceiling and at the wall above the river gallery. This network is related to the two faults that flank the limestone block in which the cave is dug. There are three types of filling. A fine scree, with bones, fed and formed by the rockfalls in the sinkhole. This scree occupies the upper part of the room. It is deposited on a blockfield with massive limestone boulders of metric size that comes from the collapsed roof. Finally, the deepest parts of the chamber are covered with a thick layer of clay that comes from the settling of the muddy water that periodically floods the lower part of the cave. There are no speleothems.

The gallery located east of the chamber is crossed by a stream that cascades into a lake. After running along the northern wall and bathing the base of the scree, the water sinks, in a diffuse way, in the scree, west of the chamber. The river is likely to have as a base level the Middle Oxfordian marls that extend below the Rauracian (Upper Oxfordian) limestone. The stream is a part of the Verneau hydrologic system (Prost et al. 2017).

2.2.3 Genesis and Stability

The bottom of the chamber is a few meters above the Oxfordian marls on which the groundwater runs. The scouring of the marls beneath the limestone beds caused their collapse. Nowadays, the water crosses the marls and reaches the lower limestone unit. The walls look stable. The only observed collapsed rocks come from the edges of the entrance sinkhole.

2.3 Bédeilhac Cave (Bédeilhac, Ariège, France)

2.3.1 Description

See Figs. 2.13, 2.14, 2.15, 2.16, 2.17, 2.18.

Due to its impressive entrance, Bédeilhac Cave has always been known by the locals. During WW2, it was used by the Germans as an aircraft engine assembly plant (Pétris 2017). Concrete slabs were installed on the floor. They are still visible. The cave is currently undergoing tourist exploitation because of the prehistoric engravings it contains. Access is from Tarascon-sur-Ariège via the D618 to Bédeilhac. A road going up on the right leads to the entrance porch of the cave (coordinates: 42° 52′ 17″ N, 1° 34′ 14″ E).

A monumental porch gives access to a large gallery with a concrete floor. Beyond the touristic part, one quickly reaches a chamber with a rock field at the end of which a

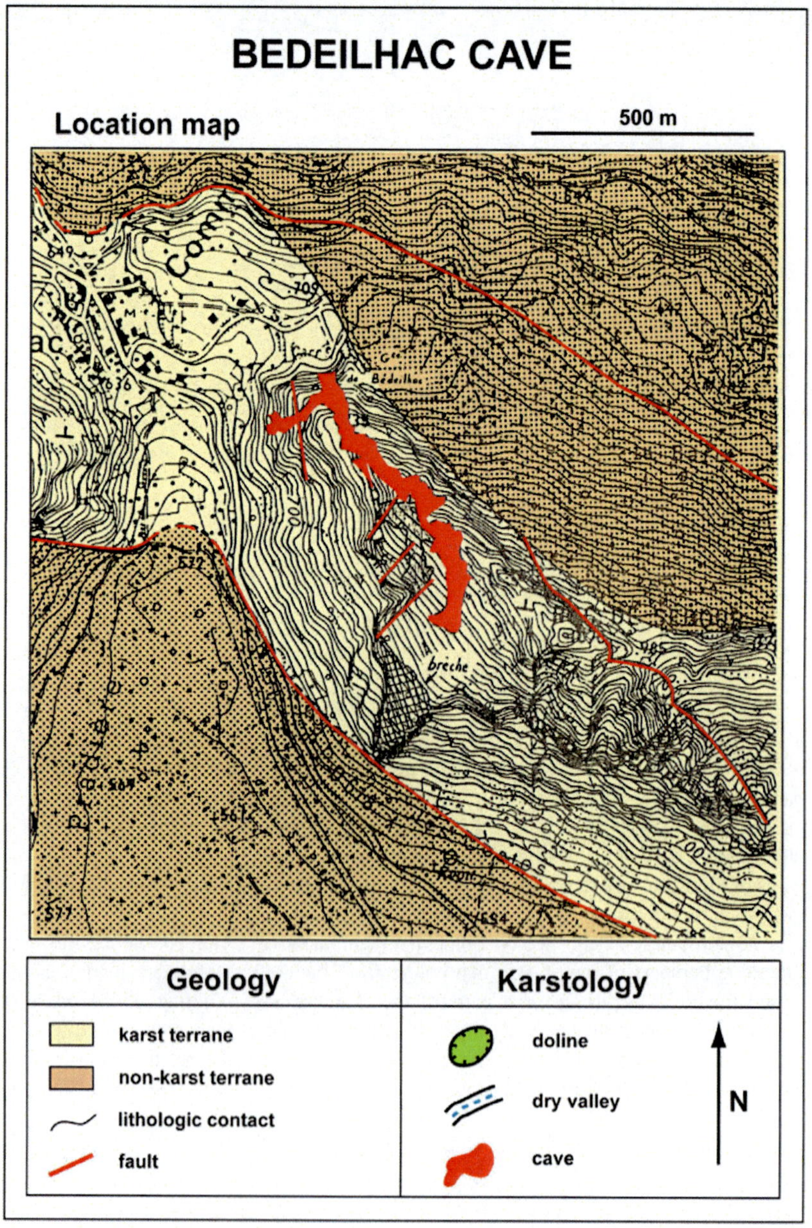

Fig. 2.13 Geomorphological map of Bédeilhac Cave

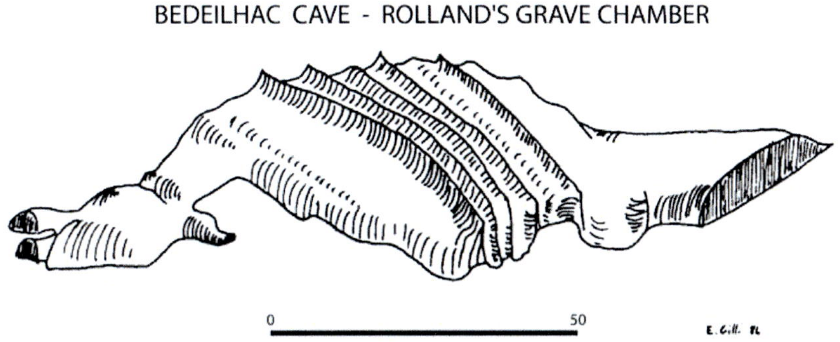

Fig. 2.14 3D sketch of Bédeilhac Cave

Fig. 2.15 Entrance of Bédeilhac Cave (photo Kvardek du. Licence Creative Common.)

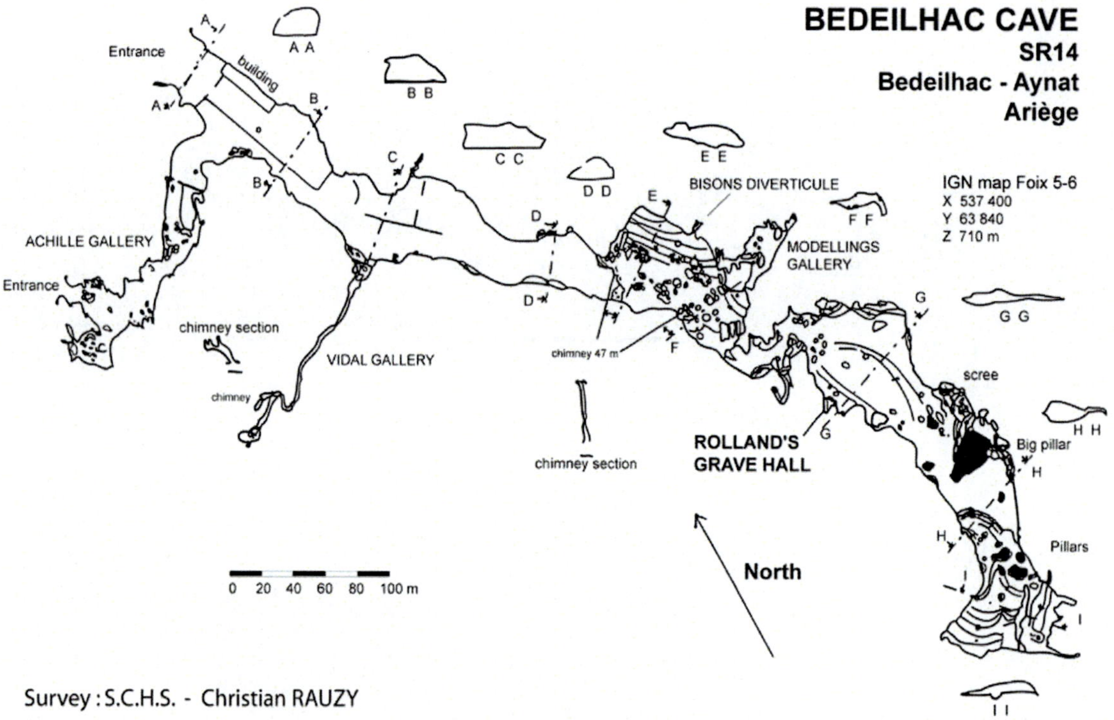

Fig. 2.16 Map of Bédeilhac Cave

gate closes a narrow passage. This is the beginning of the prehistoric parts of the cave. The floor is made of clay, the gallery widens again and forms the so-called room of the "Roland's Grave" (Figs. 2.16, 2.17) (Rauzy 1982).

This chamber is 65 m long for a more or less similar width. It has a roughly square shape. The access is through its northwest corner; the exit is at the southeast corner. The floor is relatively flat. The ceiling is a low segmental vault with many joints. Some large blocks protrude from the clay, among them is Roland's Grave, which is actually a large stalagmite lying on the floor.

Beyond this room, the gallery continues and the ceiling lowers until reaching the floor. There is no trace of current water flow. Only a channel dug in the clay at the end part of the cave testifies of the last flows.

2.3.2 Geology

Bédeilhac Cave is inside the Sédour Rock, a 1070-m-high limestone peak with steep slopes (Fig. 2.13). The elevation of the entrance is 710 m. At Roland's Grave point, the thickness of limestone above the chamber is around 280 m.

Another cave, the Pradières Cave (see Fig. 2.20), is located 150 m above Bédeilhac Cave. Both Bédeilhac and Pradières Caves are located on the western flank of the Sédour syncline, which is the northern end of the Tarascon-sur-Ariège synclinorium that is deposited, to the north, on the Arize Hercynian unit. The cave is entirely dug in the thick series of Urgo-Aptian limestone. It is a semi-crystalline gray limestone that contains numerous fossil debris. The sequence is massive, and the bedding planes are almost non-existent.

It is difficult to precisely define the geometry of the fracture system in the cave. The cracks are numerous and have varied directions. This "anarchic" fracturation is to be linked to the absence of bedding planes. Nevertheless, the N25° direction is more frequent than the other ones, with a dip to the west that varies from 25° to 75°. These discontinuities can be considered as belonging to a system of joints. It is to this system that belong most of the fractures visible in the ceiling of the room. It should also be noted that there is a fault at the chamber gate point (direction N140°, dipping west 55°) (Fig. 2.17).

The clay soil is deposited on a complex cave filling that contains crystalline pebbles of glacial origin (Sorriaux 1982). The room which is before the main chamber is

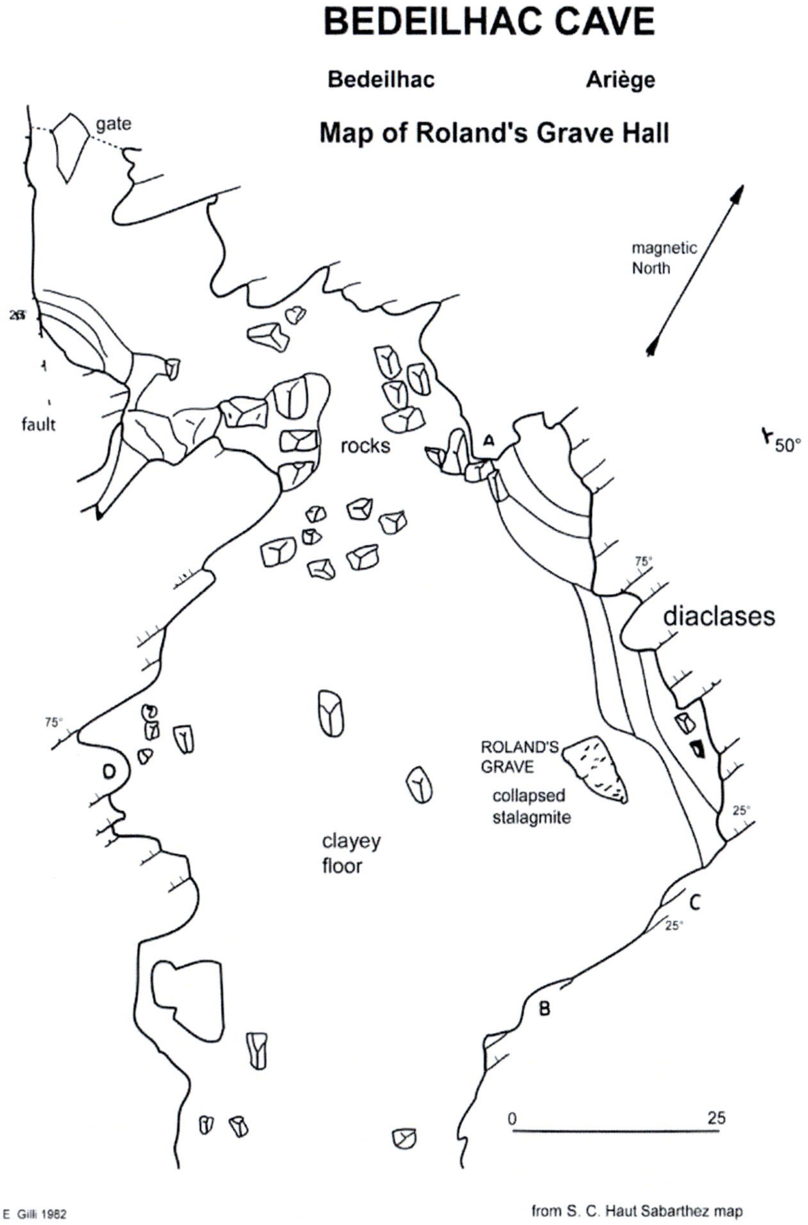

Fig. 2.17 Map of Roland's Grave Hall

occupied in its central area by a block field whose elements come from the collapse of the ceiling. On the other hand, it is important to point out, because the fact is rare, the absence of fallen blocks in Roland's Grave chamber. The absence of collapsed blocks and the shape of the ceiling suggest a digging by dissolution and mechanical erosion (Fig. 2.18). Despite the flat shape of the ceiling vault, this chamber seems to be stable for a long time.

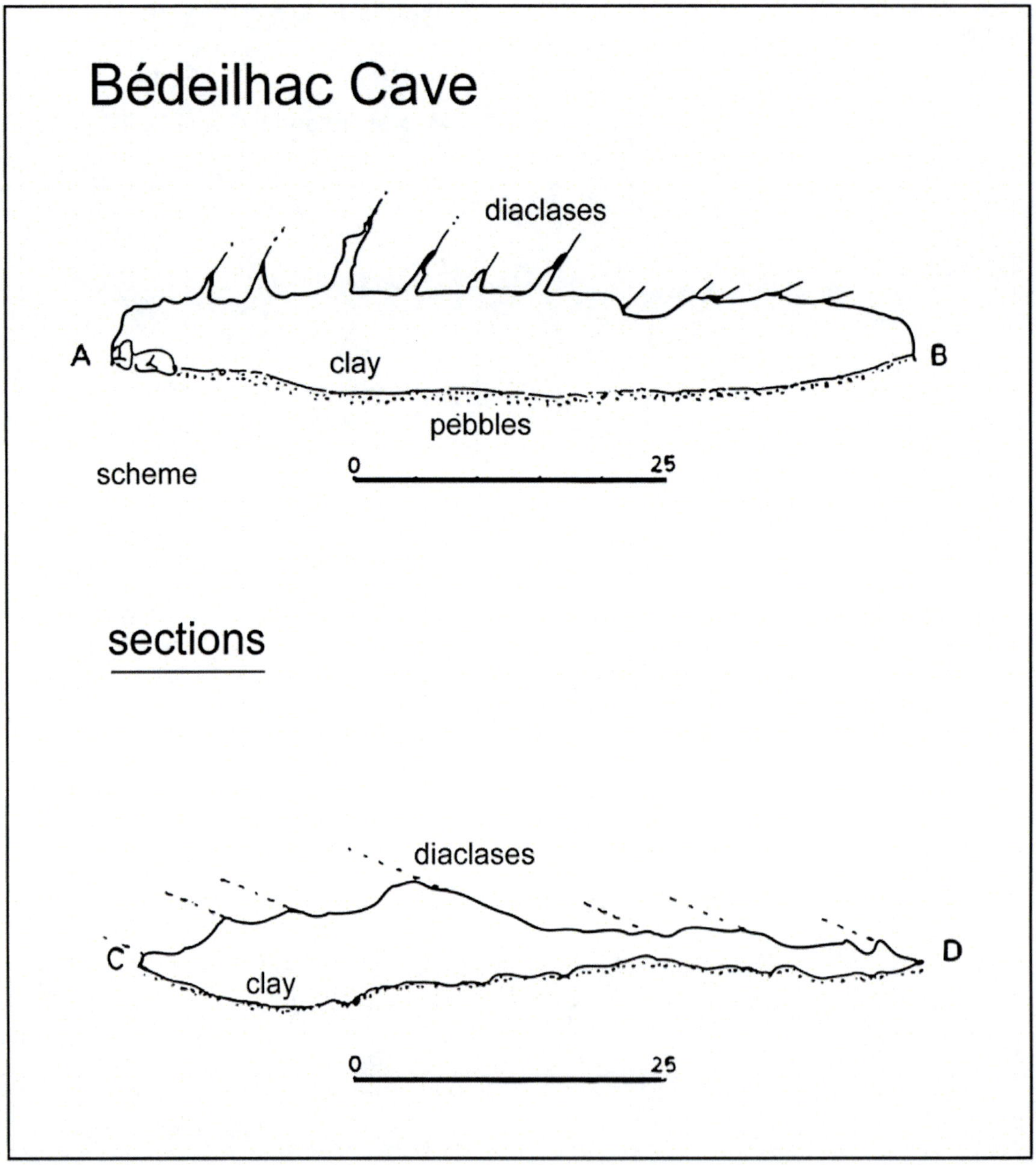

Fig. 2.18 Cross sections A–B and B–C of Roland's Grave Chamber

2.4 Bournillon (Chatelus, Isère, France)

2.4.1 Description

The cave is highly visible in the landscape from the Bourne Valley and therefore obviously known since forever (Figs. 2.19, 2.20). Moreover, the cave is easy to reach. On the D531 from Villard-de-Lans to Pont-en-Royans, before Choranche village takes a small road that leads to a hydro-electric plant. From the plant, a path goes toward the cliff on the left bank of the Bournillon riverbed. The entrance is at the end of the path (Coordinates: 45° 03′ 16″ N, 5° 25′ 57″ E).

The cave begins with a huge porch 105 m high, 60 m wide and 160 m long. The right part is occupied by a scree. The Bournillon River runs along the left wall. An artificial dam was built at the entrance of the cave to store the water. The reservoir lake extends in the center of the porch. Water is collected there to feed the power plant (Figs. 2.21, 2.22, 2.23, 2.24, 2.25, 2.26, 2.27, 2.28, 2.29, 2.30, 2.31). At the end of the porch is the departure of a cave network explored on several kilometers (Caillaut et al. 1999). A large dry gallery whose entrance is located in the western part of the porch, at the top of the scree slope, makes it possible to reach the underground river. The ceiling has a characteristic shape of a reversed staircase (Fig. 2.19). A river runs through the cave. Its discharge may reach 30–40 m^3/s during floods. It is used to supply a hydroelectric plant.

2.4.2 Geology

The Bournillon Cave is located on the left bank of the Bourne Valley, 400 m above sea level, in a vast circus in the Urgonian cliffs. These 470-m-high limestone cliffs lay on an impermeable marly substratum. Bournillon underground river flows at the contact between the limestone and the marls. The cave network extends below the St. Julien-en-Vercors karst plateau which constitutes its

Fig. 2.19 Bournillon Porch

Fig. 2.20 Measuring the ceiling of Bournillon Porch with a helium balloon

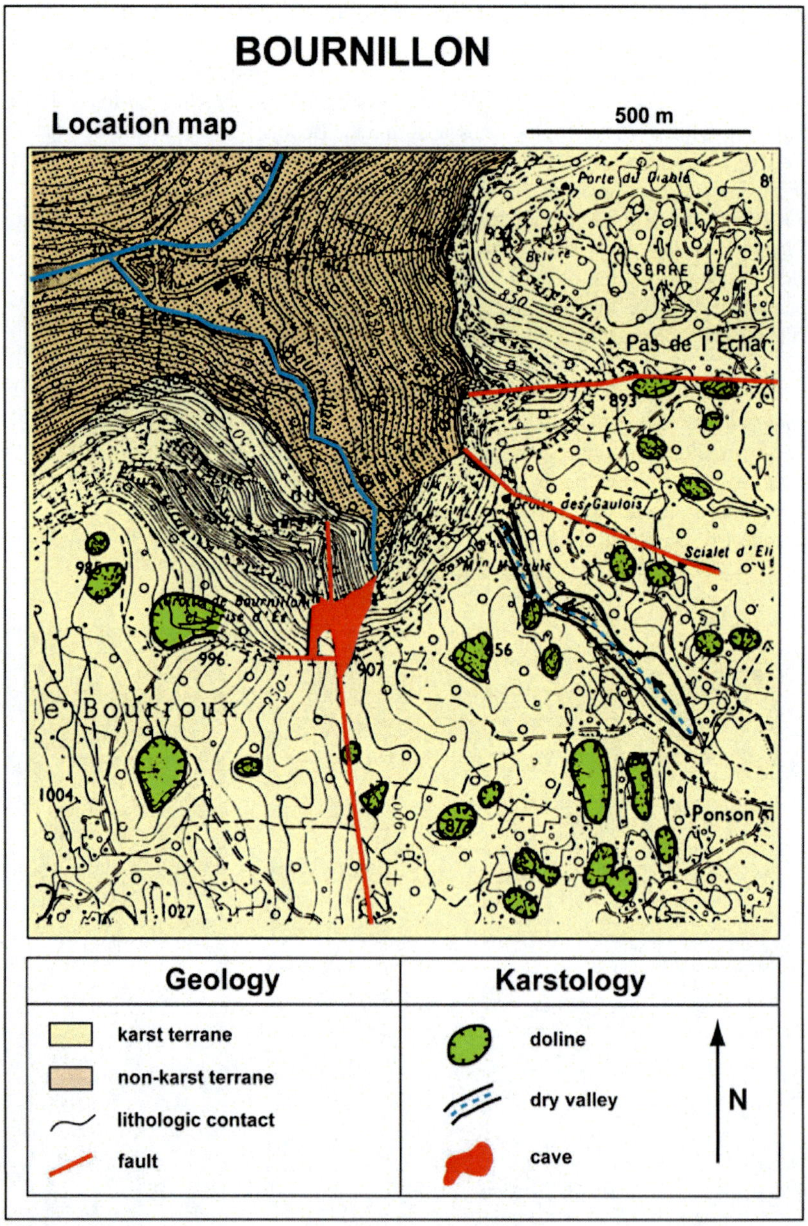

Fig. 2.21 Geomorphological map of Bournillon Cave

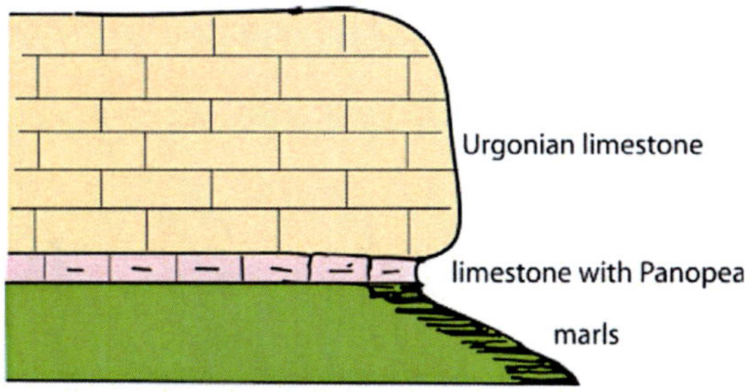

Fig. 2.22 Geological section (from Moret 1950)

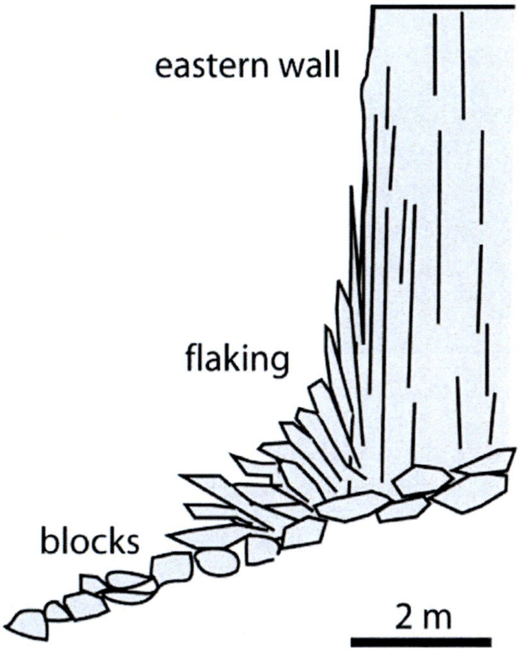

Fig. 2.23 Flaking on the western wall of Bournillon porch

catchment area. The thickness of the limestone above the porch exceeds 350 m (Fig. 2.21).

The cave is dug in Cretaceous tabular series, at the base of high Urgonian limestone cliffs. The groundwater level is controlled by the Hauterivian marly-limestone layers which are visible along the path up to the cave entrance. The contact is underlined by a spectacular overhang (Fig. 2.22). At its base, the Urgonian is a sandy limestone that contains Panopea. It is stratified in decimetric beds. Above, the limestone becomes more massive with banks of decametric thickness. They are separated by more soluble interbeds; however, the morphological aspect is that of a very thick and resistant limestone unit.

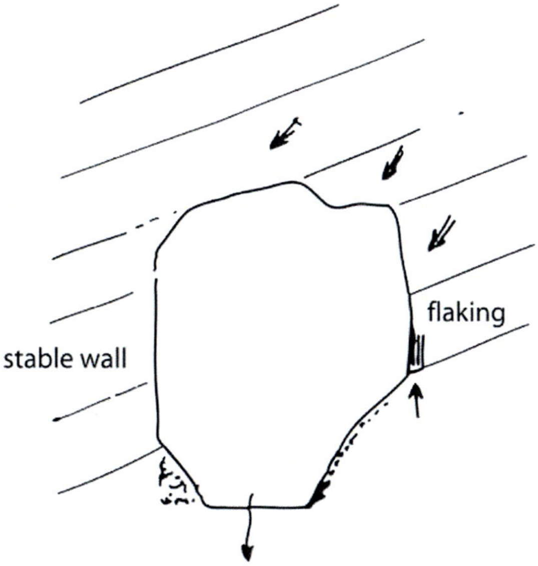

Fig. 2.24 Stress oriented by the dip in Bournillon porch

It is this massive sequence that allowed the development of the ceiling of the Bournillon porch and gave it its appearance in reversed staircase comparable to a corbelled vault. The beds have an N20° strike and a 15° dip to the east. The cavity is aligned on a large N0° joint. The western wall of the porch is the eastern wall of this crack (Fig. 2.27).

The floor is covered with blocks of variable size (decimetric to metric). The outer edge of the entrance slope is earthy and shaped by the water that pours from the cliff. At the back of the porch, the scree is thinner, and the blocks rarely exceed 20 cm in diameter. They come from a very spalled zone which explains their size. The river cleared off the scree all over the eastern part of the porch. There is no speleothem on the ceiling.

2.4.3 Genesis and Stability

The high dimension of the porch is related to several factors:

- the presence of a large N0° joint;
- the decompression of the limestone near the slope;

Fig. 2.25 Natural pillar close to the upstream gallery

- the erosion of the Hauterivian marly substratum that created a significant overhang comparable to that seen on the path to the cave.

Important flaking phenomena are observed at the western wall in relation to the presence of the large N 0° joint (Fig. 2.23). This instability may be related to the dip (Fig. 2.24). Some limestone beds seem more susceptible to flaking than other ones (Renault 1984, oral communication).

Moving away from the entrance, toward the interior of the massif, the phenomena of flaking become even more spectacular. They are always located on the west wall of the porch. At the end of the porch, to the west of the wet gallery, there is a natural pillar with spectacular chipping. In the main upper dry gallery (Fig. 2.25), the flaking phenomena are not as important as in the entrance porch. One can note the presence of a large joint whose wall forms the western side of the gallery.

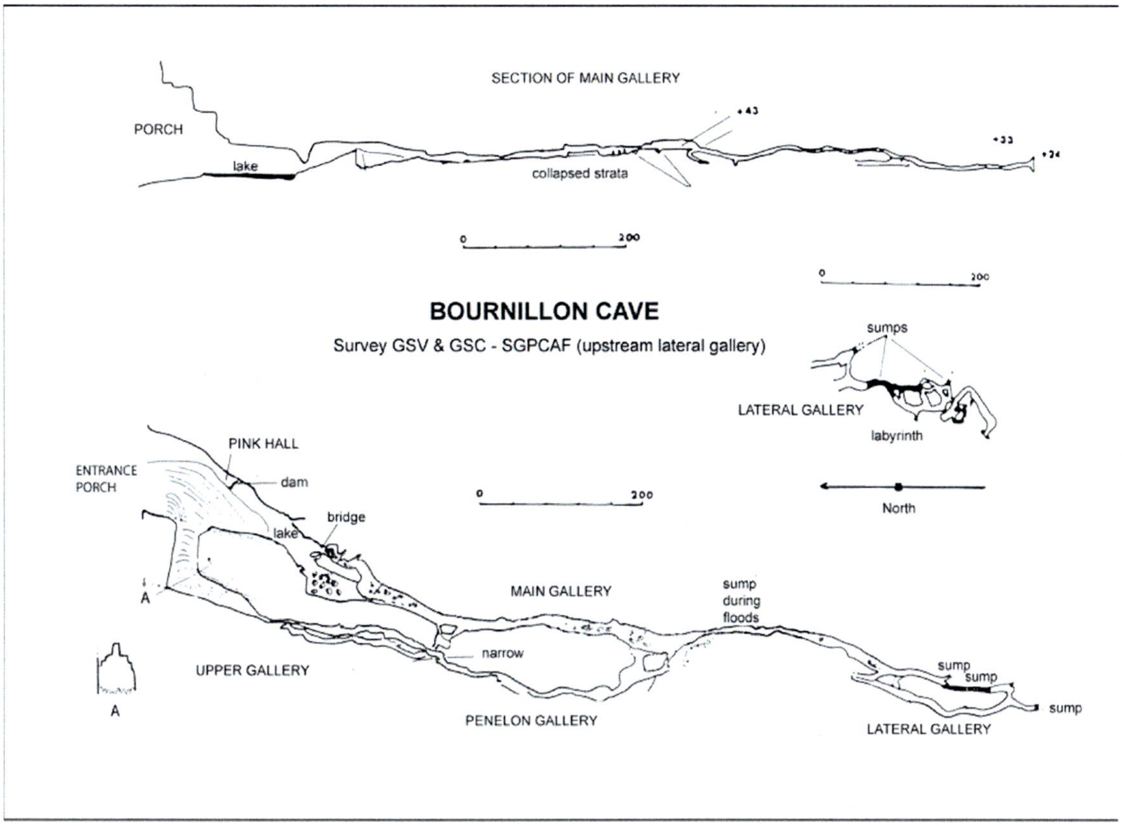

Fig. 2.26 Survey of Bournillon Cave

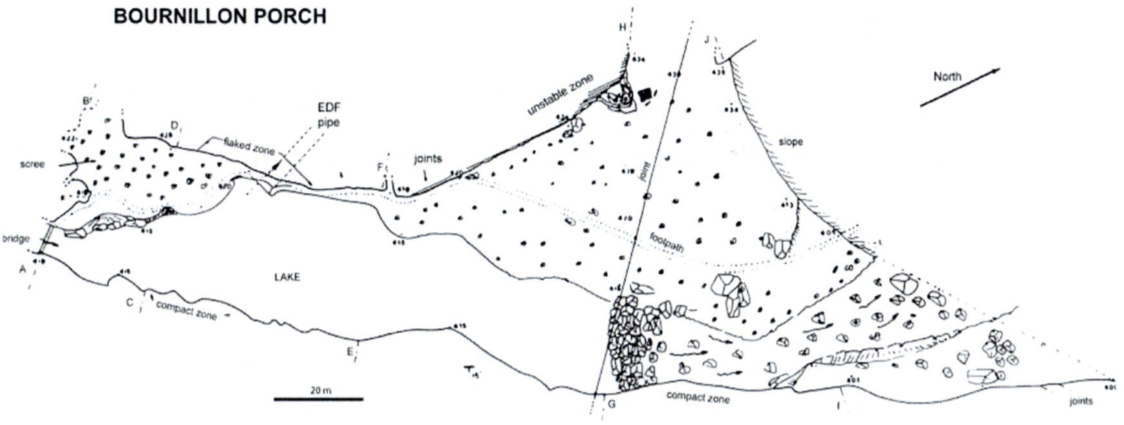

Fig. 2.27 Map of Bournillon Porch

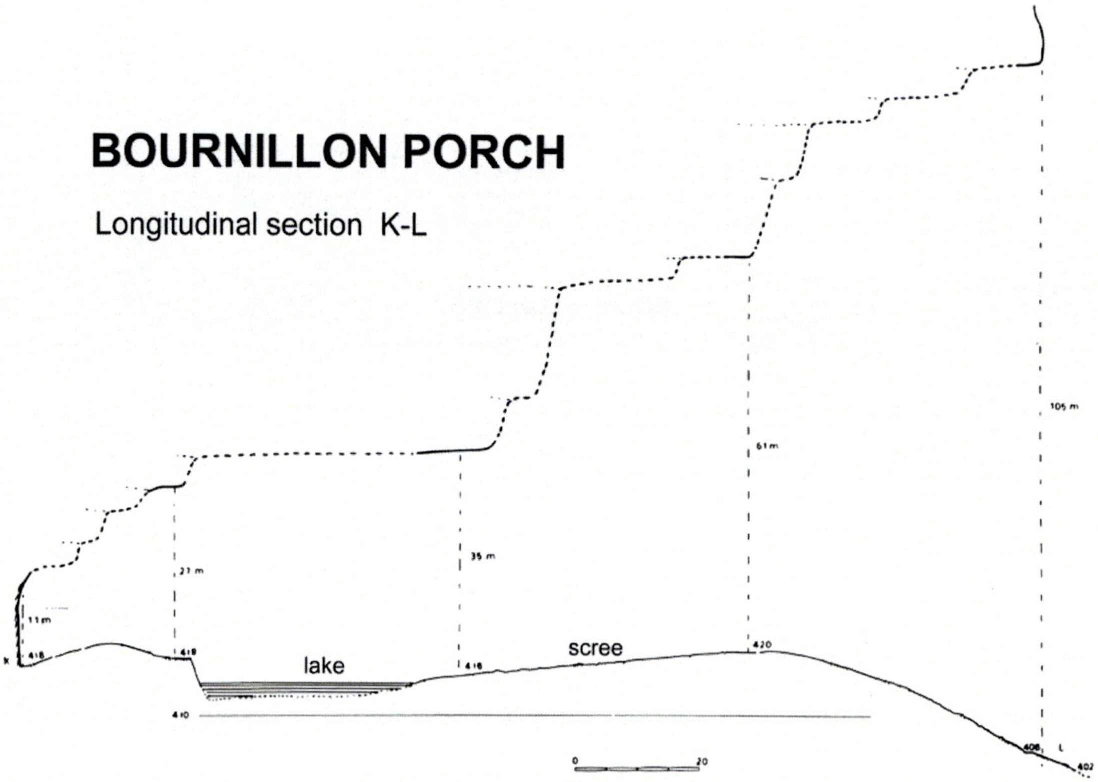

Fig. 2.28 Longitudinal section of Bournillon Porch

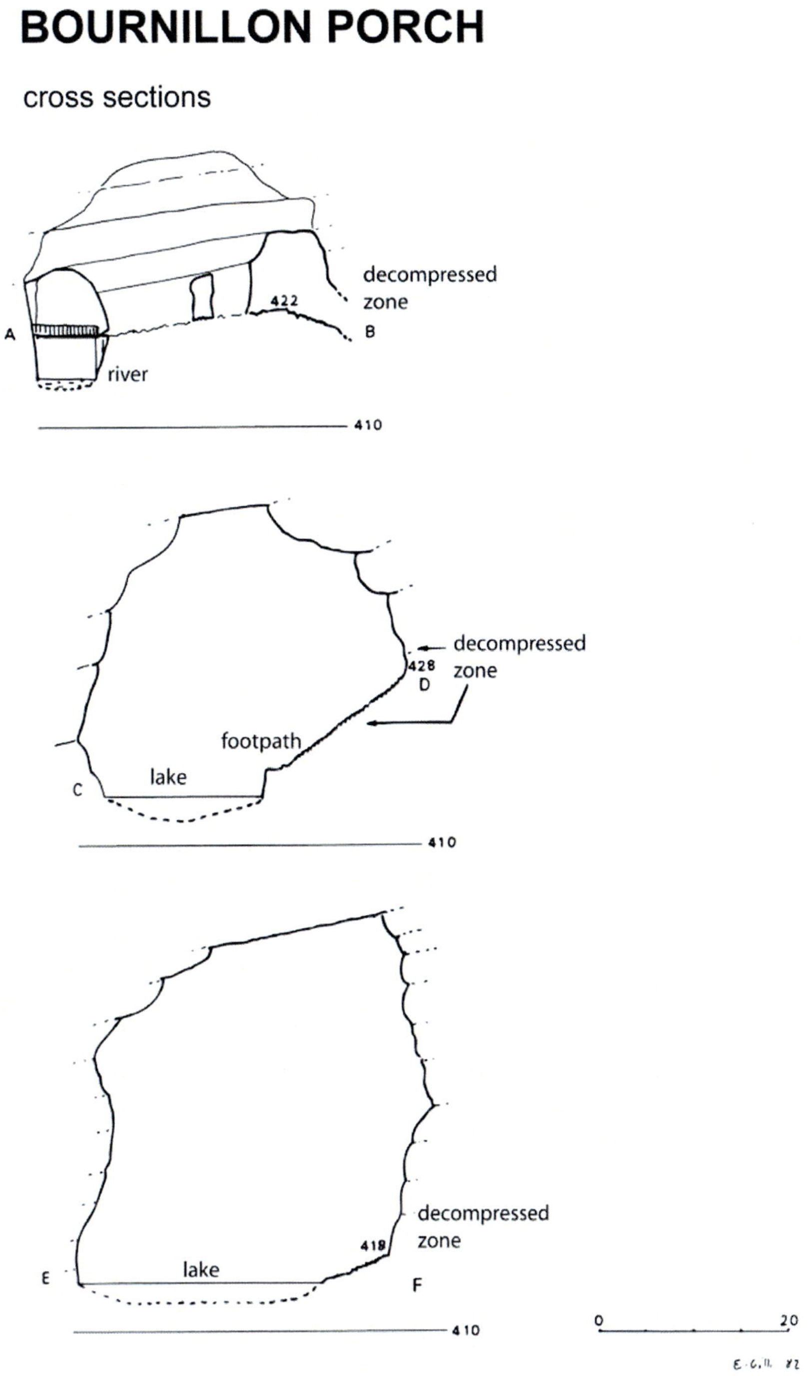

Fig. 2.29 Cross sections A–B. C–D. E–F of Bournillon Porch

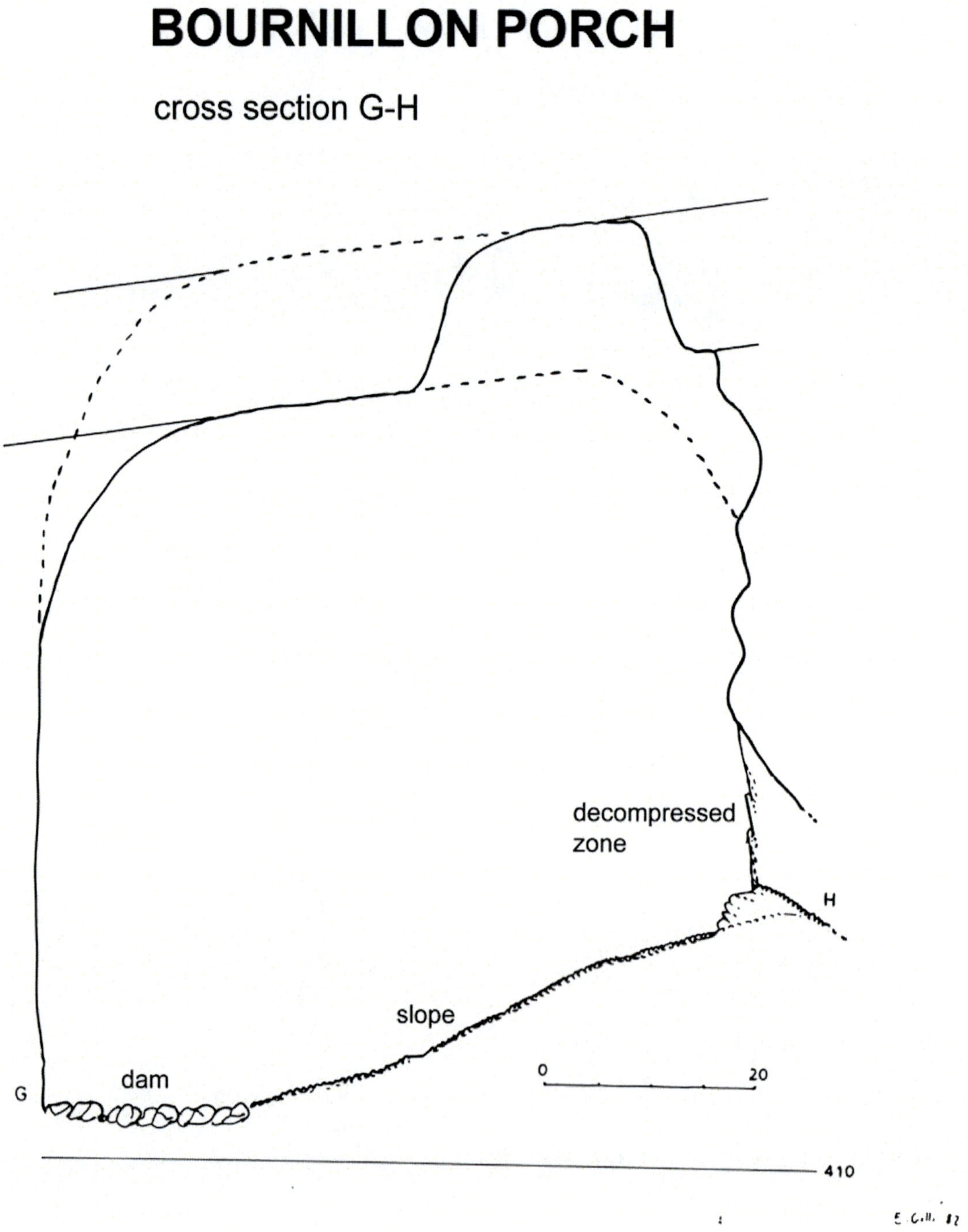

Fig. 2.30 Cross section G-H of Bournillon Porch

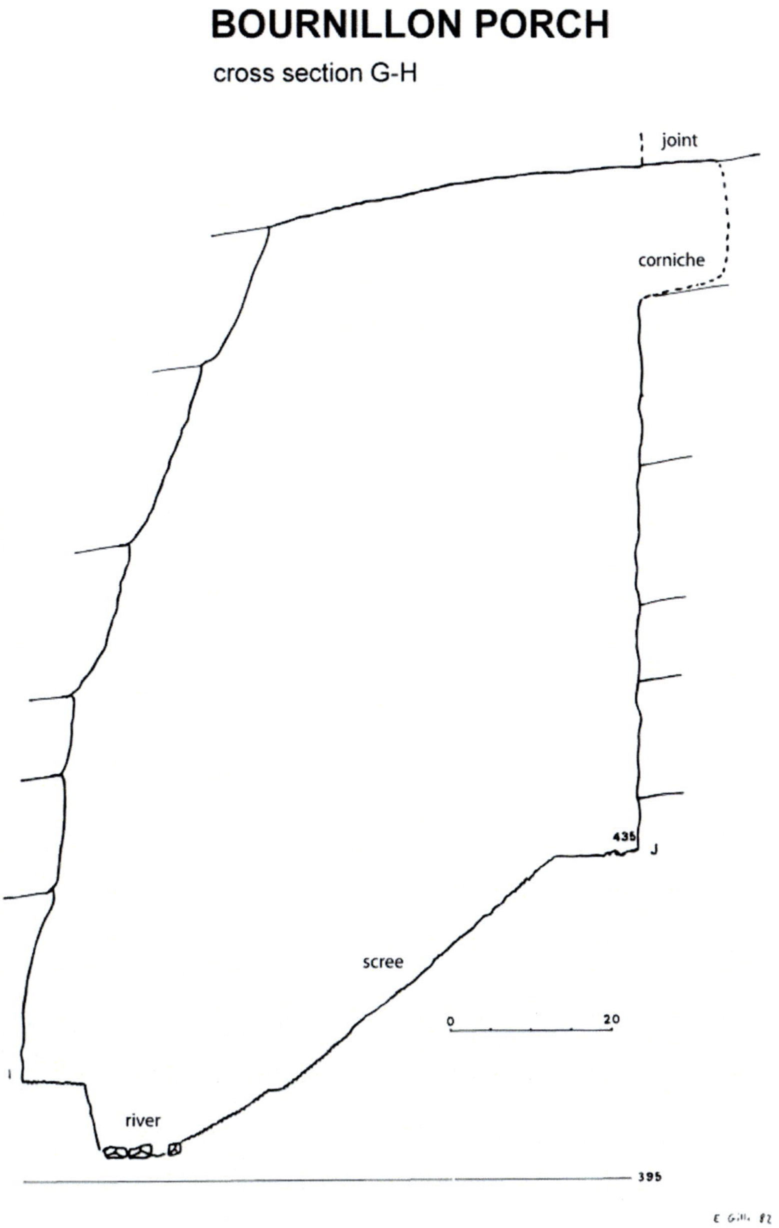

Fig. 2.31 Cross section I–J of Bournillon Porch

2.5 Castelette (Nans-Les-Pins, Var, France)

2.5.1 Description

See Figs. 2.32, 2.33, 2.34, 2.35, 2.36, 2.37, 2.38, 2.39.

From Brignoles, drive on the N7, then the D1 and the D80 to Nans-les-Pins. From there a road goes up to the Sainte Baume. Before arriving on the plateau, leave the car on a well visible parking area that is located on the right side of the road in a bend. A path descends to the foot of the limestone. It is steep and little marked except in its lower part where it goes up to the left until the natural entrance to the cave. The artificial pit dug by cavers is located about 50 m in height on the left. The access is by a small but clearly visible path (coordinates: 43° 20' 37.05 N; 5° 45' 28.38 E).

The hand-dug pitch gives access to a narrow conduit that leads to the top of a small pitch opposite of which is the entrance of a small room. On the right side of this room, a 25 m pitch arrives in a vast gallery where a river flows. Go up the river to a first sump that can be shorted by an upper passage and then arrive at a second sump, which is actually a duck where one has to immerse.

The gallery continues until a blockfield which fills the lower part of the room. By crawling between the blocks, one ends up in the chamber. It is 110 m long and 55 m wide. It is an elongated room, with a very low ceiling and a blockfield floor that rises sharply northward until it reaches the ceiling. Among the blocks, there are large slabs collapsed from the ceiling (Fig. 2.33). The blockfield is affected by subsidence along the southern wall and at the western end. It is caused by the river that runs below this part.

The cave is traversed by an underground river that passes under the room. The water comes from a small river that drains the marly slope at the foothill of the Sainte Baume cliffs. It is swallowed by the ponor of Plan d'Aups Plateau. The underground stream exits at the Castelette springs and forms a tributary of the Huveaune River.

Fig. 2.32 General view of Castelette Chamber

Fig. 2.33 Collapse of thin limestone beds from the ceiling

2.5.2 Geology and Hydrogeology

The natural entrance of the cave is a resurgence that gives access to a gallery which quickly ends on a sump. The hand-dug well has shorted this sump. The resurgence is 590 m above sea level at the foot of a thick limestone ridge, in a cirque, upstream of the Huveaune Valley. Above the ridges, a horizontal karst plateau extends at an altitude of 670 m. This plateau is bordered on the south by marl embankments topped by thick limestone ridges—the Sainte Baume Cliff. In its limestone part, the plateau is very karstified, with many karrens that emphasize the fracturing. A ponor, the Tourne, is located southwest of the cave. Castelette chamber is located 150 m from the north of the plateau; the thickness of limestone is around 30 m.

The cave is dug in the Upper Jurassic limestone and dolomite series of the Plan d'Aups Plateau. To the south, this unit is overthrusted by the Sainte Baume series. To the north, east-west faults limit a collapsed compartment. On the surface of the Plan-d'Aups Plateau, the limestone is intensely fractured according to a double network of vertical joints oriented E-W and N-S. The rock consists of 5-cm-thick beds grouped in large beds 0.5 to 1 m thick. These beds have a weak cohesion between them. The dip of the layers is 20° to the southwest (Fig. 2.35). Bedding-plane slip is important, and numerous slickensides are observed. The direction of the movement is N 0°.

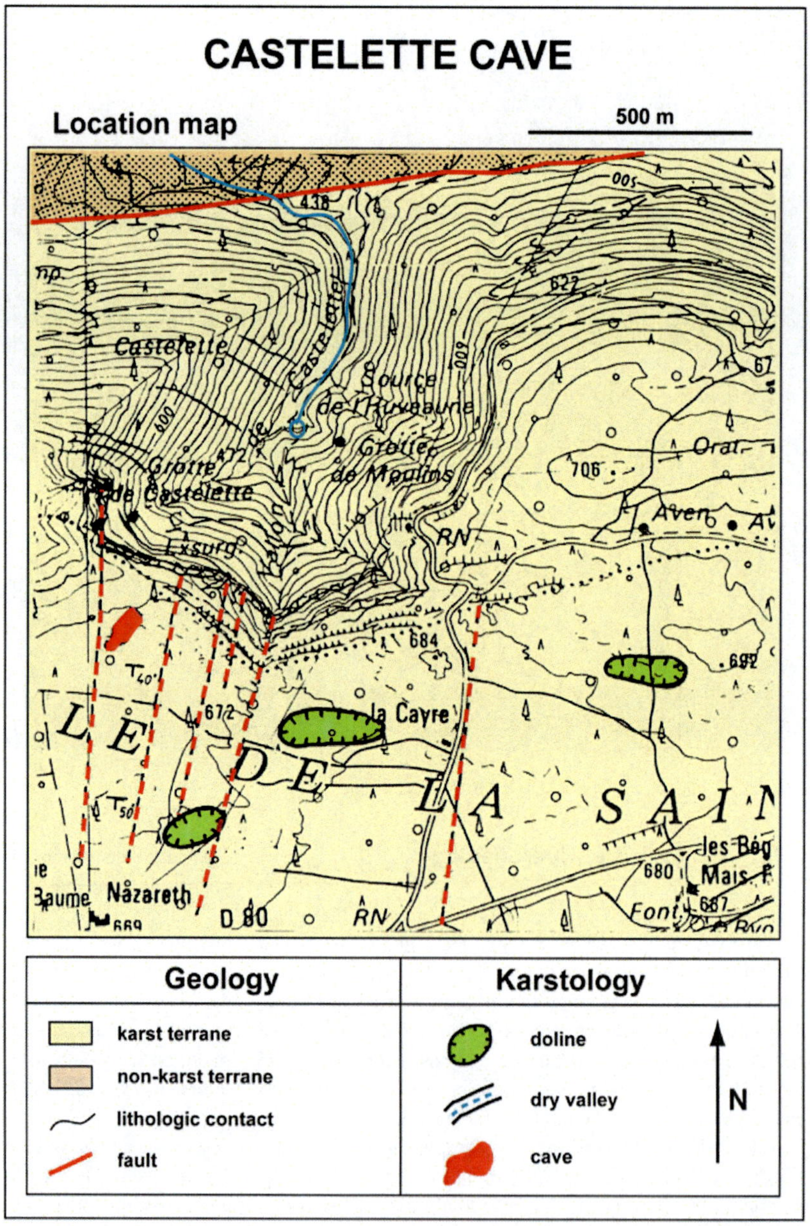

Fig. 2.34 Geomorphological map of Castelette Cave

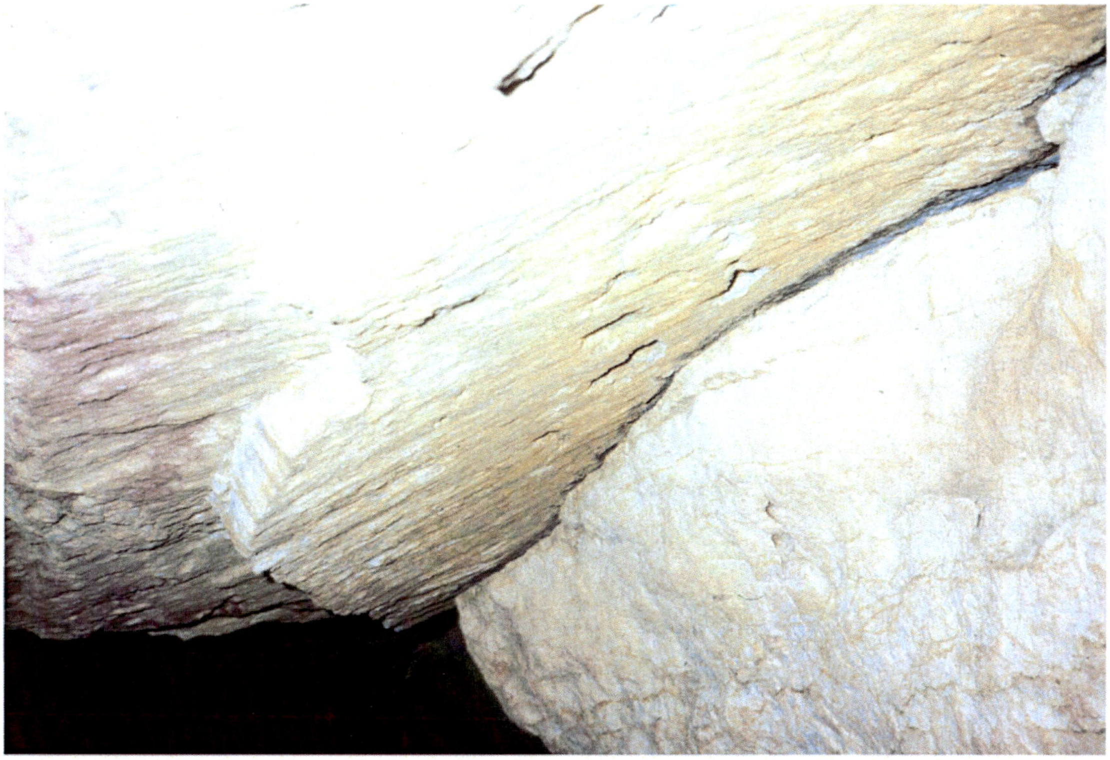

Fig. 2.35 Bedding-plane slips

Fig. 2.36 3D sketch of Castelette Chamber

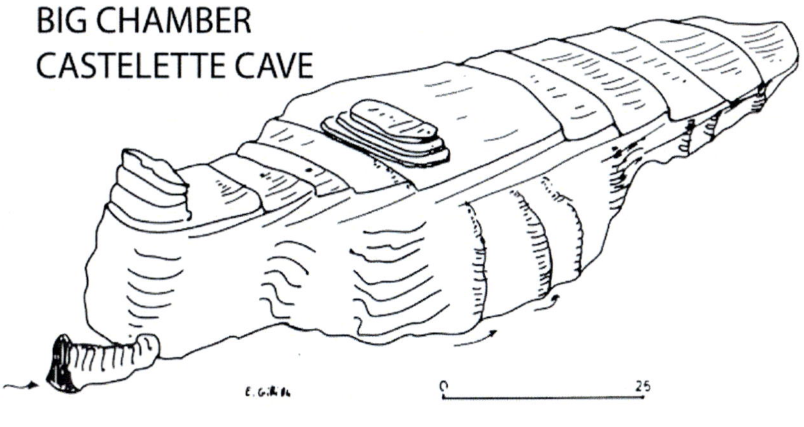

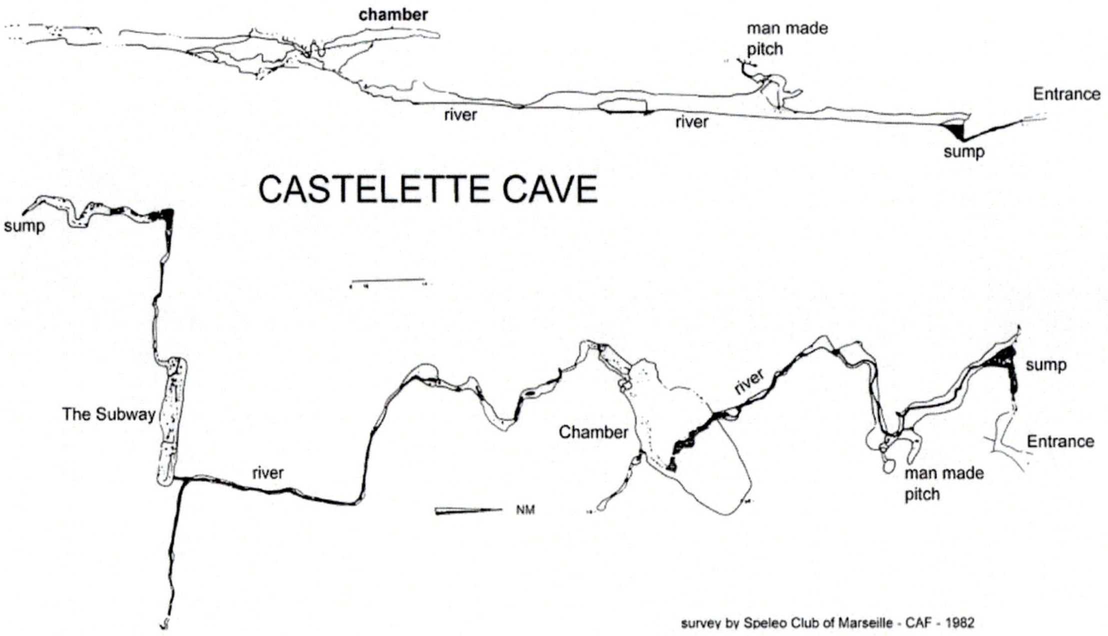

Fig. 2.37 Survey of Castelette Cave

Despite the bayonet layout of the galleries that are presumably oriented by the rock discontinuities, it is difficult to recognize the fracture directions in the chamber. Only a vertical diaclase of orientation N10° is clearly visible in the southern part. The most important visible discontinuities are therefore the bedding-plane slips. The floor consists entirely of blocks and slabs. The upper blocks appear to come from recent collapses. In some areas, it is possible to easily drop down slabs from the ceiling. Speleothems are rare and are concentrated at the northern end of the chamber. The southern and southwestern parts that are close to the river are covered with clay.

2.5.3 Genesis and Stability

It is a collapse room. Despite the absence of sand in the room, it is conceivable that the cavity originated by emptying a pocket of dolomitic sand. It was then enlarged by the collapse of the ceiling at the points of weakness formed by the slipped interbeds. The instability signs are numerous, not only at the level of the ceiling, but also in the blockfield on the floor. The racking of the scree by the river seems to be active as shown by the orientation of the slabs and the steep slopes of the floor above the stream.

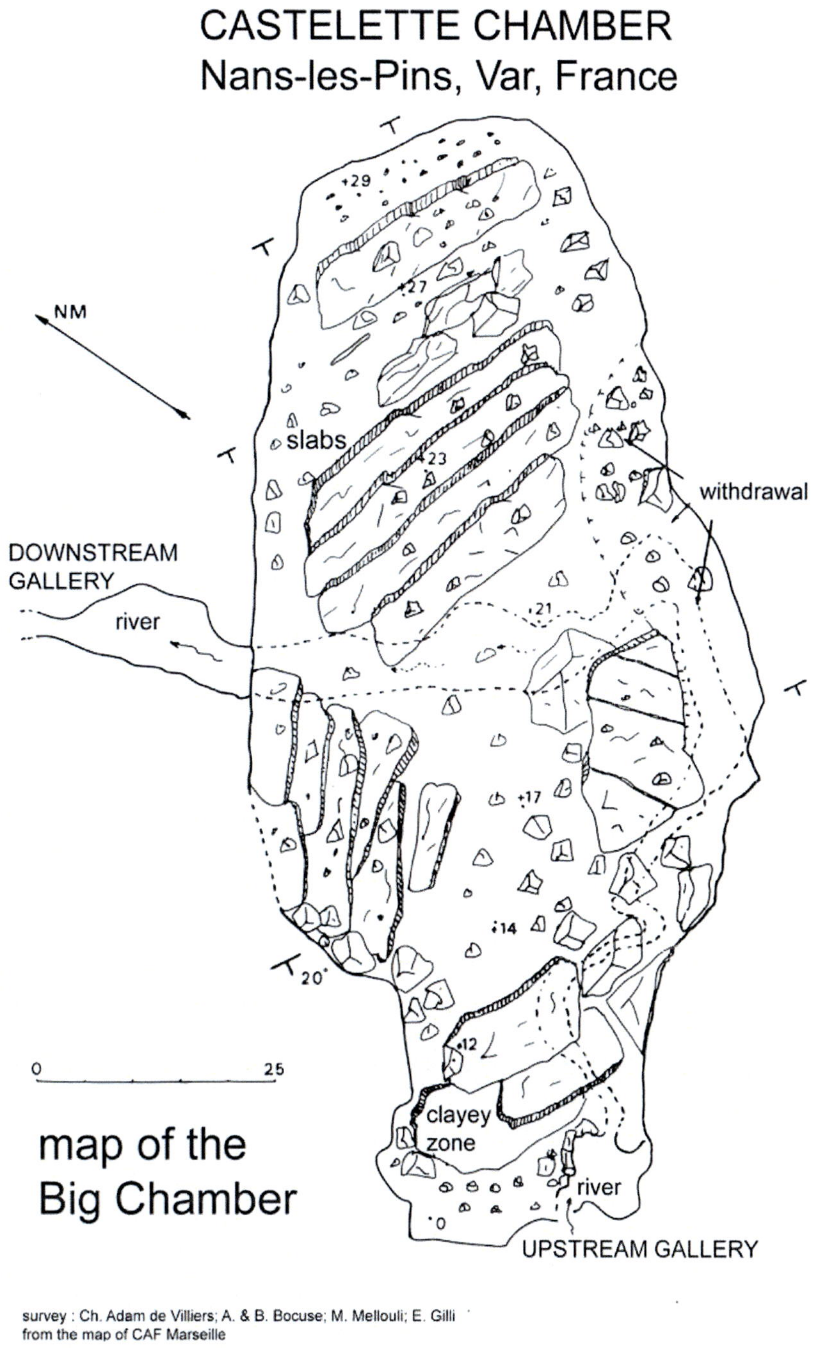

Fig. 2.38 Map of Big Chamber in Castelette Cave

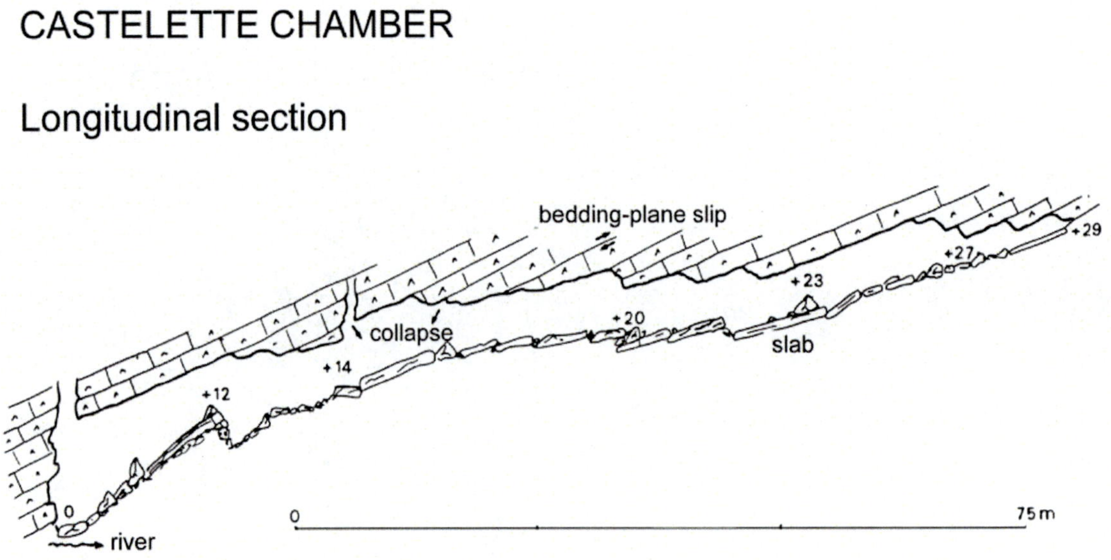

Fig. 2.39 Longitudinal cross section of Big Chamber

2.6 Cennet and Cehennem Chasms (Narlikuyu, Silifke, Turkey)

2.6.1 Description

See Figs. 2.40, 2.41, 2.42, 2.43, 2.44, 2.45, 2.46, 2.47 from Aygen (1970).

In southern Turkey, Cennet and Cehennem Chasms can be reached on the coastal road from Antalya to Mersin, 20 km east of Silifke (Fig. 2.40). Both chasms were described by Aygen (1970). They are in a tourist site which is well indicated and accessible with an asphalt road.

2.6.1.1 Cennet Düdeni (Paradise Chasm)

This is a large collapse sinkhole with vertical edges. It measures 275 × 125 m. The floor is inclined to the south and converted into crop terraces. A paved path, from Byzantine era, leads to an antique chapel. It is located at the entrance of a huge gallery (approximately 50 m wide) that leads to an underground river. The water comes from a scree and sinks at the bottom of the cave, where it joins the major drain currently impenetrable. Indeed, it flows into a flattener, 15–20 cm high, about 30 m above sea level.

2.6.1.2 Cehennem Düdeni (Hell Chasm)

This cave is very different from the previous one, which is located only 75 m away from it. It is a collapse shaft similar to Padirac Shaft. Its depth varies between 80 m and 100 m. The entrance is approximately 50–55 m in diameter; it is horizontal and opens in the middle of the karren, in massive limestone (Fig. 2.41). The bottom of the pitch is occupied by an earthy scree slope and much of it is overgrown. In the northern part, a large sloping gallery leads to the bottom of the chasm (120 m). It is not possible to reach the river in Hell chasm.

2.6.2 Geology and Hydrogeology

Both caves are located on the southern flank of a hill, 150 m. a.s.l., 1 km north of the shore of the Mediterranean Sea. The landscape is a coastal karst type. The karst landforms are well developed. The limestone often appears bare, with well-developed karrens. The soil is gathered in terraces and forms a few cultivable areas.

Surface hydrology is absent. The thickness of limestone above the large gallery of Cennet Chasm is about 50 m. In Cehennem Chasm, there is an important overhang that presumably corresponds to a vestige of the cave ceiling before its collapse (Figs. 2.42, 2.46). There, the thickness of the rock is only a few meters. However, it is possible that this overhang is a secondary form that was shaped after the collapse of the ceiling. Likewise, the dissolution of the surface karren could continue after the collapse of the ceiling and thus reduce the thickness of the limestone above the cavity.

The rock is a microfractured white limestone, which is often brecciated with a red cement. Dip is 4° toward the east, and the strike is N0°. The limestone beds are very massive;

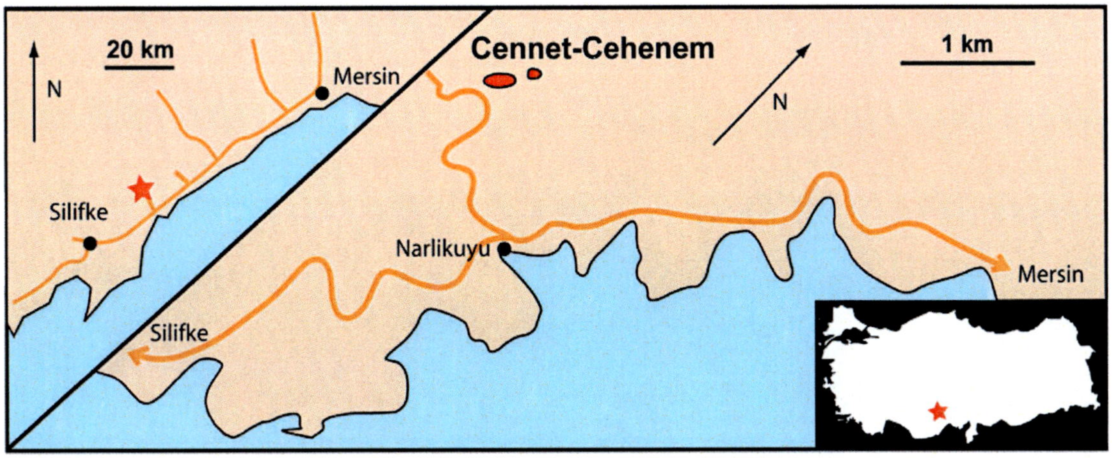

Fig. 2.40 Location of Paradise and Hell chasms

Fig. 2.41 Christian church in Cennet (Paradise) Shaft (photo Korkut Tas—Creative Common—Panoramio)

Fig. 2.42 128-m-deep Cehennem (Hell) Shaft. Note the remnant of the ceiling in the background (photo Korkut Tas—Creative Common—Panoramio)

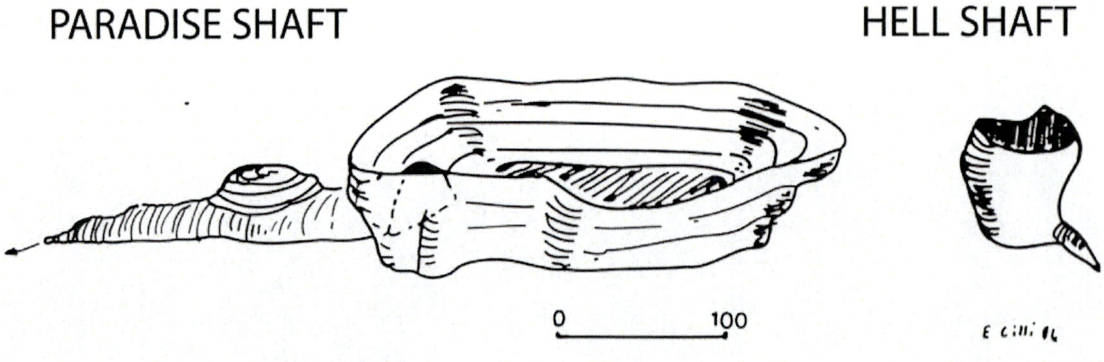

Fig. 2.43 3D sketch of Cennet and Cehennem chasms

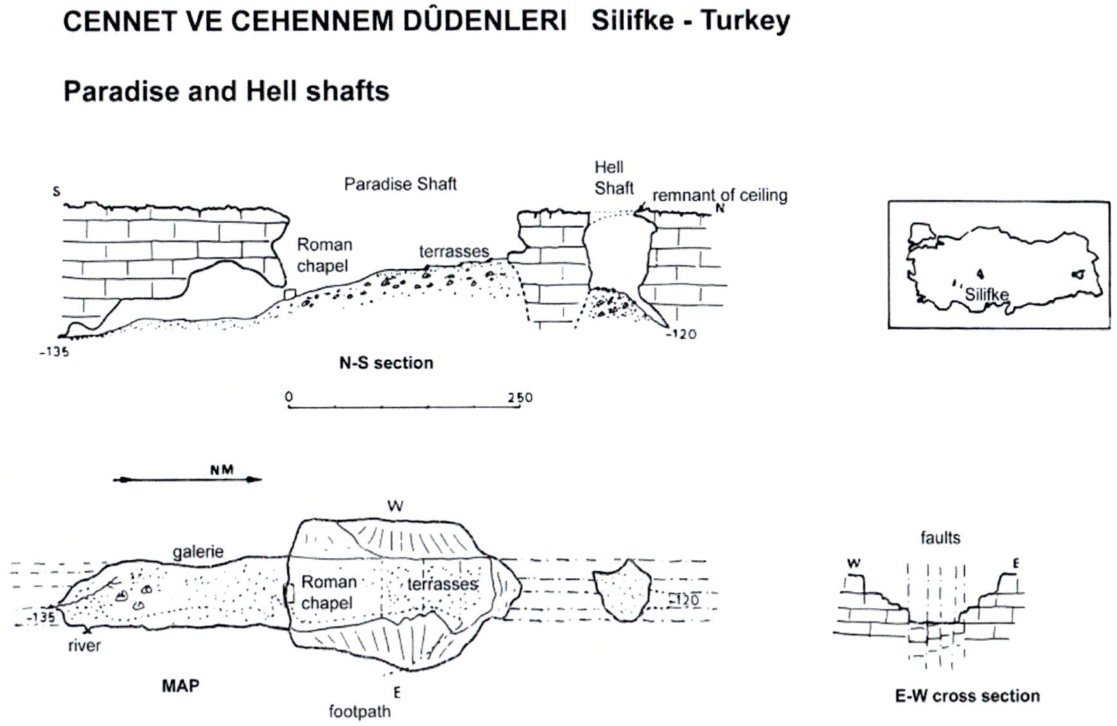

Fig. 2.44 Survey sketches of Cennet and Cehennem Shafts

their thickness varies between 1 and 10 m. They form a very compact series. This limestone accepts large spans as evidenced by the vestige of the ceiling in the Cehennem Chasm. The average thickness is only 5 m; and the span exceeds 20 m (see N-S section), which is despite the importance of surface lapies.

The two caves are aligned on a vertical network of N170° faults. The east and west walls of the galleries of the chasms are faults belonging to that network. The morphology of the two cavities is therefore very affected by these fractures (Fig. 2.44). In their collapsed parts, both chasms are occupied by blockfields and earth in which the vegetation has grown. The land was even terraced in Byzantine times. The filling of the gallery of the Cennet Chasm becomes more and more clayey toward the bottom. At the end part of the river, the floor consists of rolled pebbles. It is therefore possible to observe the removal of the cave filling by the underground river.

The resurgence of the river that flows in the Cennet Chasm is below the sea level, in the small bay of Narlikuyu, 1.2 km from the sinkholes. The cave network seems to develop in the interbeds following the dip. The slope is weak (2.5%). The discharge of the river was not measured.

2.6.3 Genesis and Stability

Obviously, Cennet and Cehennem Chasms correspond to the collapse of the roof of a giant gallery aligned on a system of parallel N170° faults. The emptying of the crushed limestone between the joints generated a vacuum in which there is a remnant at the gallery of Cennet Chasm. The collapse of the ceiling obstructed this ancient gallery, forcing the water to flow in a currently impenetrable passage on a bedding joint. In Cennet Chasm at the level of the passage of the network of faults, the rock is very dissociated; many blocks appear in equilibrium. However, the current stability of the cavity does not seem to be affected. In Cehennem Chasm, the gallery looks very stable. Although it is partly crossed by the river, the scree does not show subsidence signs. The Byzantine chapel is perfectly preserved as well as the other buildings (stairways, terraces) built on the cave filling.

Fig. 2.45 Map of Hell Shaft

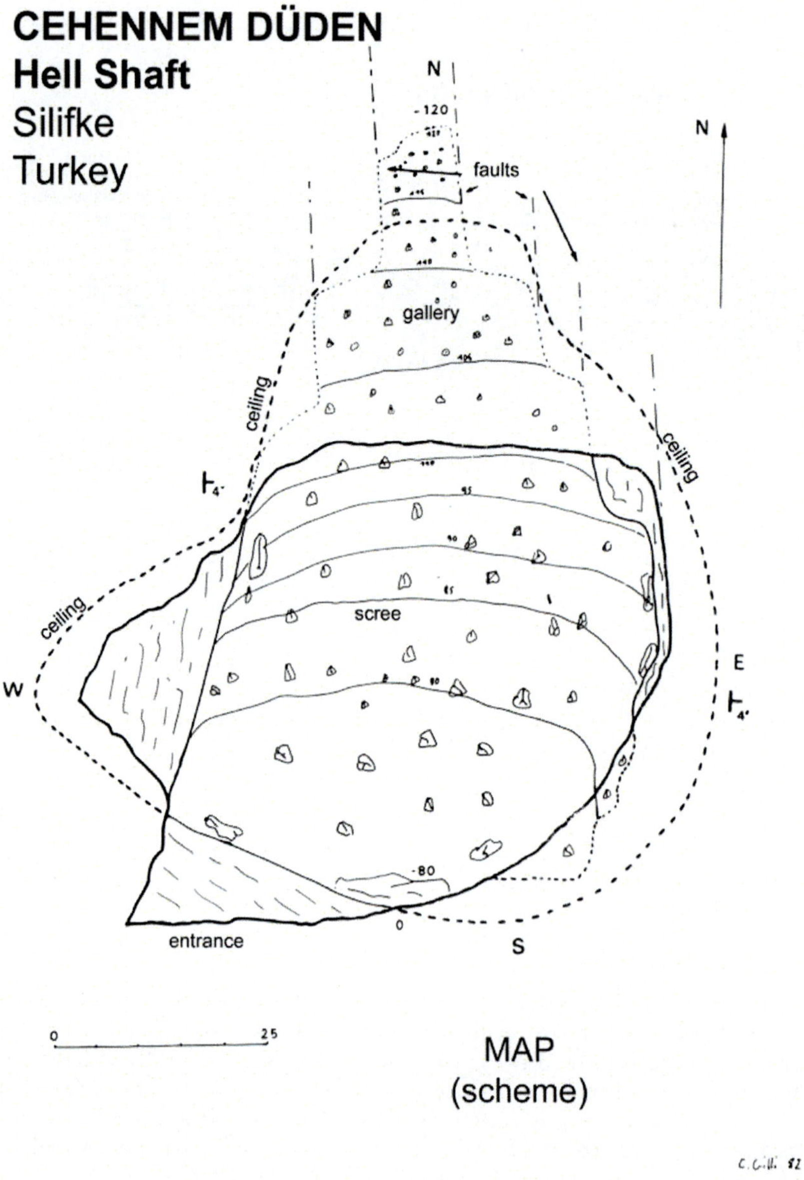

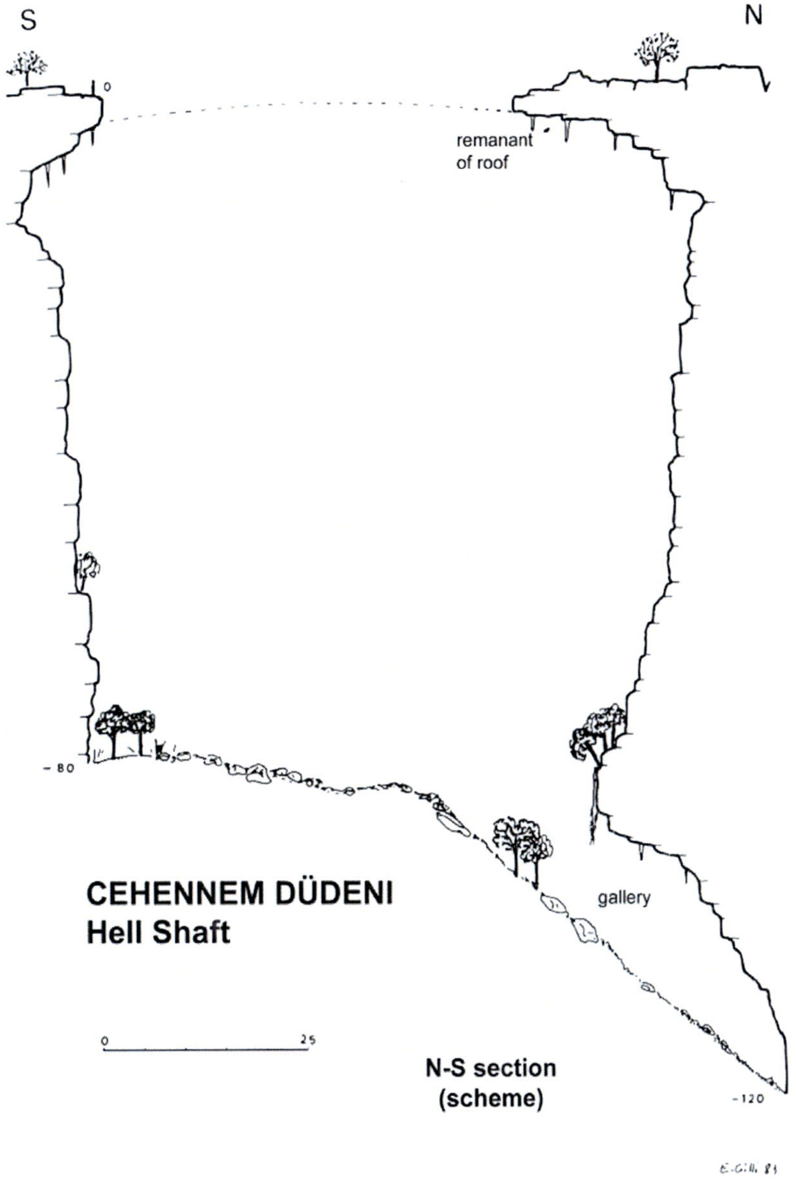

Fig. 2.46 N-S cross section of Hell Shaft

Fig. 2.47 E-W cross section of Hell Shaft

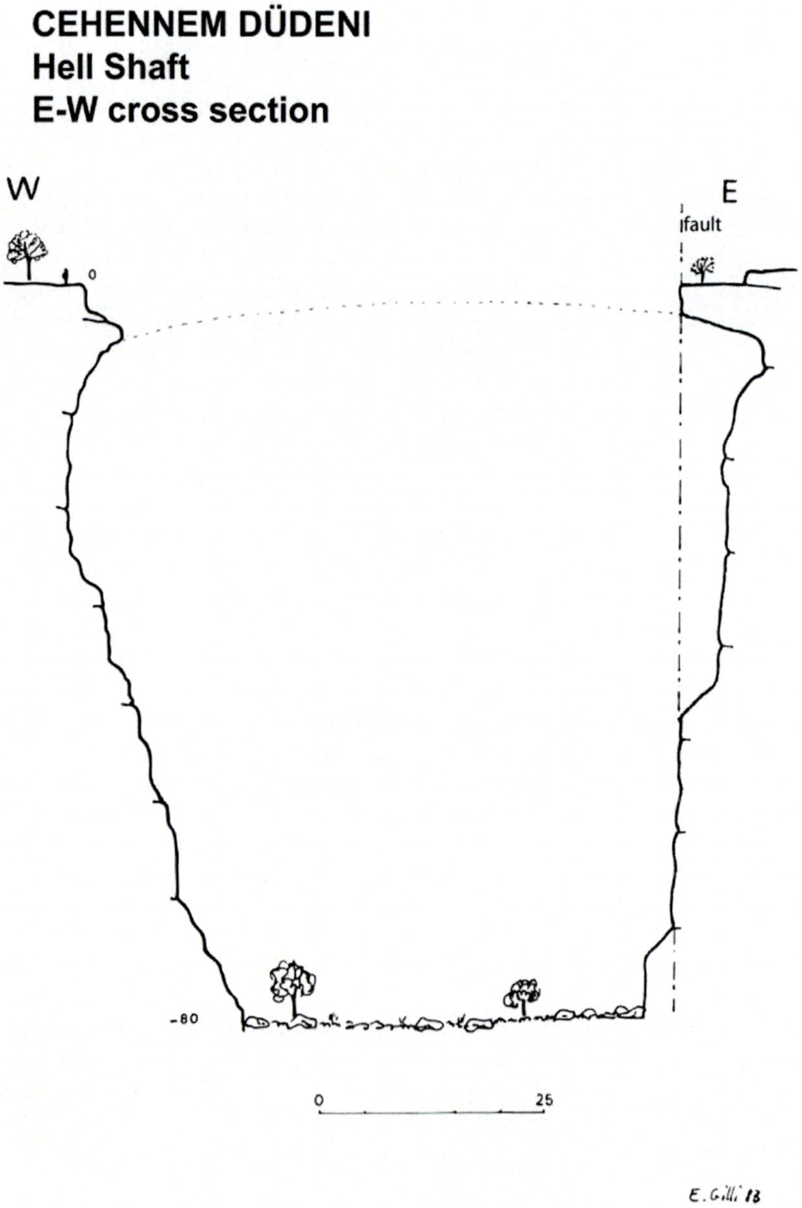

2.7 Mont Blanc Hall—Champclos Fountain (Les Vans, Ardèche, France)

2.7.1 Description

See Figs. 2.48, 2.49, 2.50, 2.51, 2.52, 2.53.

From Les Vans, drive toward the west on the D901, then the D408 to Naves hamlet. A hundred meters before the first houses, a road goes down to the bottom of the valley. A path starts from an old mill and follows the river; it is marked by chevron patterns. It passes close to the natural entrance of the cave: the Fontaine de Champclos spring. Then climb along a small valley until a man-dug pitch is reached that allows easy access to the chamber (coordinates: 44° 23’46.0201” N; 4°6’13.0201” E). The cave was surveyed by Chabaud and Rivol (1976).

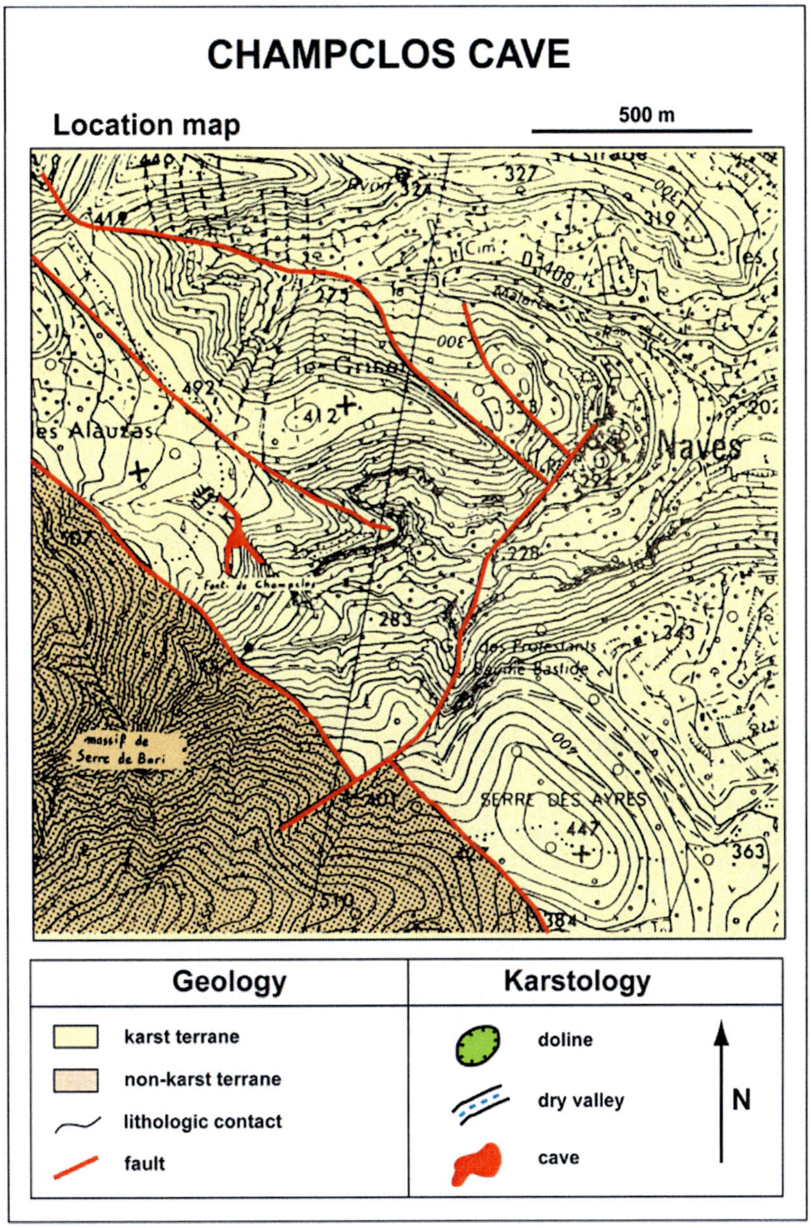

Fig. 2.48 Geomorphological map of Champclos Cave

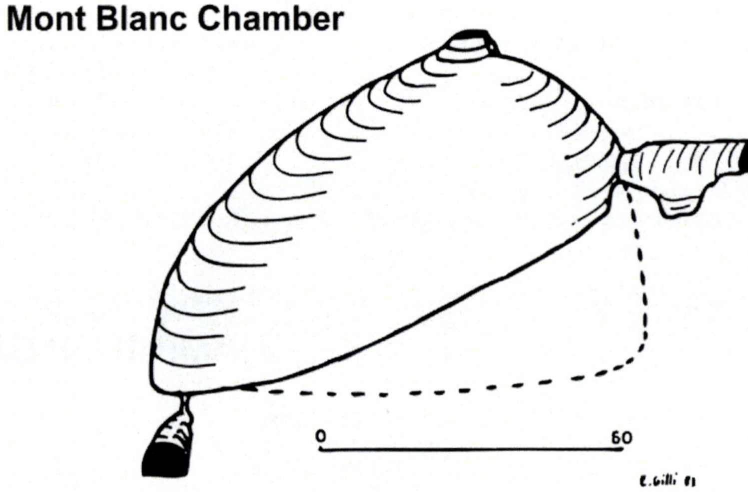

Fig. 2.49 3D sketch of Mont Blanc Chamber

The Mont Blanc Hall is triangular in shape and measures 45 m wide by 50 m long. The southern part is occupied by a chamber covered with flowstone scree slope inclined to the southeast. A gallery opens in each corner of the room:

- north, upstream, where it joins the wet cave network;
- southwest, downstream to the spring;
- southeast, to an upper dry gallery where the man-dug pitch.

The walls and ceiling form a regular rounded bell. The limestone beds are well individualized. The river that runs through the cave passes under the scree in the northeastern part of the chamber.

2.7.2 Geology and Hydrogeology

The chamber is located under a valley that borders the small Alauzas Plateau (elevation: 450 m), where various carbonate formations are present. The slopes are ravined by erosion. They collect the rainfall of the massif of Serre de Bari (elevation: 910 m) which overlooks the plateau to the west. The spring which is the natural entrance to the cave is 310 m.a.s.l. The hall is located 320 m from the spring.

The thickness of limestone above the ceiling is 55 m. The chamber is relatively close to the slope; it is about 80 m from the south valley (Fig. 2.48). The cave is dug in Jurassic sedimentary formations. The bedding is horizontal. To the west, this series is in fault contact (Banne-Villefort Fault), with the primary metamorphic terranes of the Serre de Bari. This fault is a normal dextral one oriented N130°. This direction is also observed on the plateau of Alauzas where there is also a relatively dense set of N10° joints.

The chamber is dug in an alternation of limestones and marly limestones. The beds are 20 cm thick; they are separated by layers of laminated and calcified marls several centimeters thick. Above the ceiling, the series consists of thick limestone beds clearly visible in the man-dug pitch entrance of the cave. These formations are from Lower Oxfordian. The limestone layers are horizontal. Despite the proximity of the fault and the bayonet aspect of the cave network, no fracture is visible in the room.

Along the northwestern wall, the floor is sandy. The rest of the room is occupied by a blockfield covered with flowstone. Flowstone is more and more important toward the dry gallery; it completely cements the scree. A calcite coating covers the walls and hardens the marly layers.

The cave is crossed by a small river whose discharge varies between 1 and 100 L/s. It passes under the Mont Blanc room along the northwest wall. The exit of the water is the Fontaine de Champclos spring which feeds a tributary of the Bourdaric River. The presence of metamorphic pebbles in the riverbed shows that the water comes from the massif of Serre de Bari.

2.7.3 Genesis and Stability

The Mont Blanc Hall is located at the junction of two galleries and results from the clearing by the river of marly soft levels; the arched ceiling enlarged until reaching a massive limestone bed under which it has been wedged. The river now flows under the room and appears to no longer participate in the digging. The walls of the room are covered with

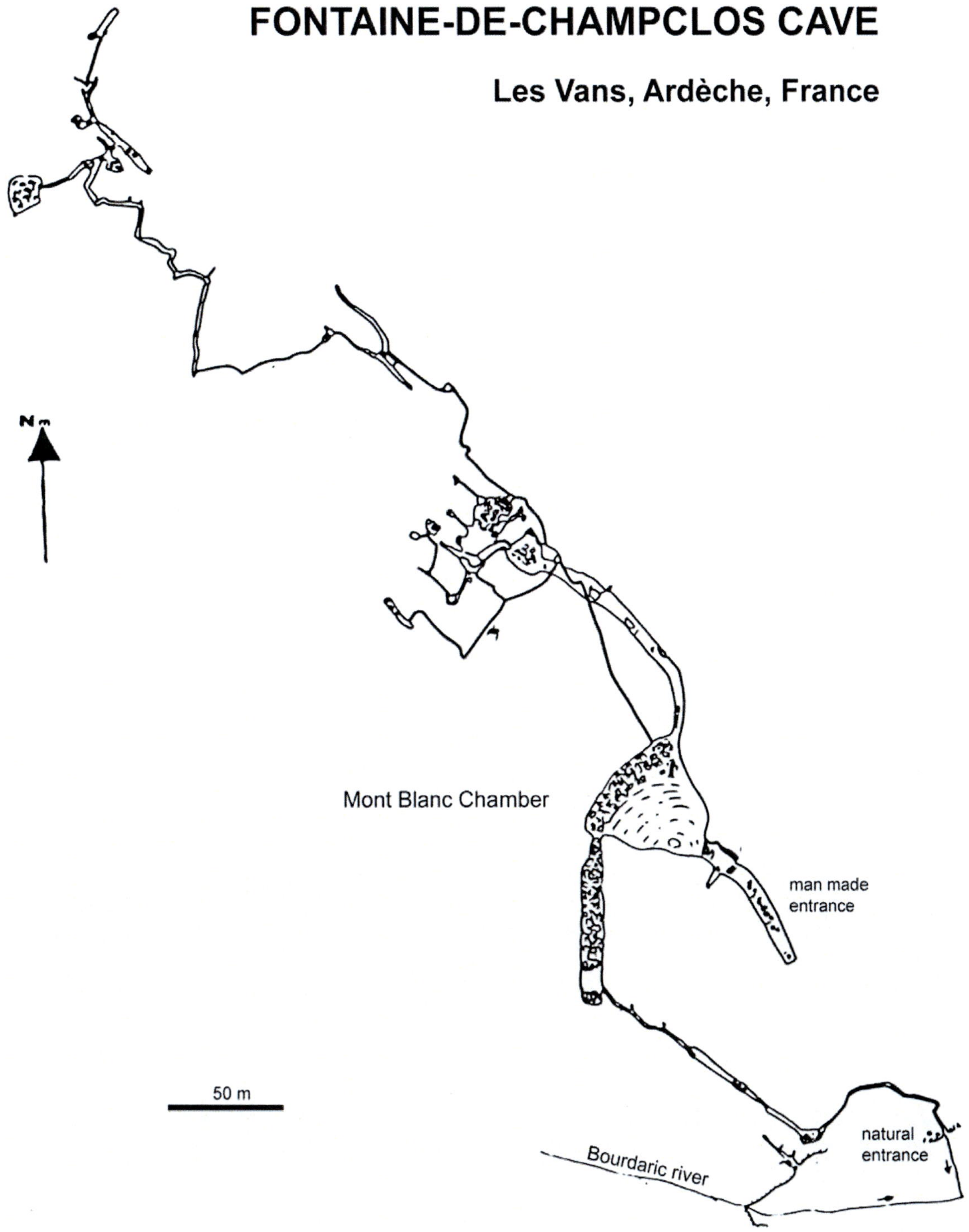

Fig. 2.50 Map of Champclos Cave

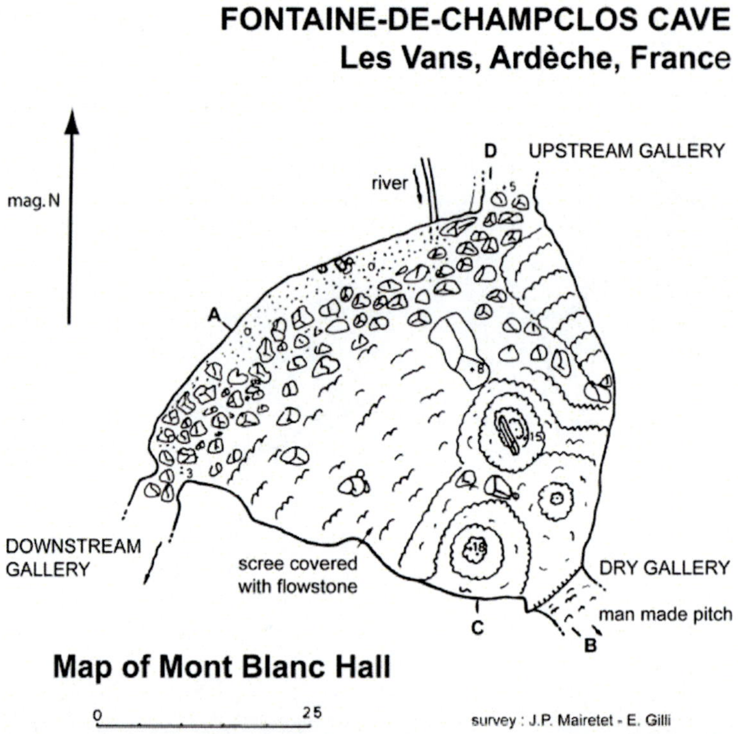

Fig. 2.51 Map of Mont Blanc Hall in Champclos Cave

Fig. 2.52 Mont Blanc Chamber (Photo Ph. Crochet)

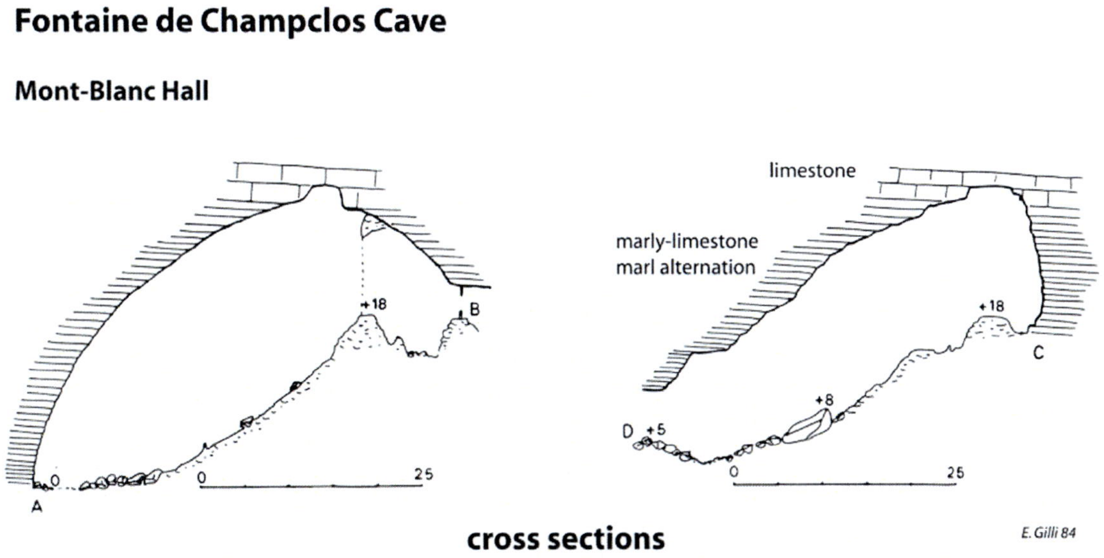

Fig. 2.53 Cross sections of Mont Blanc Hall

a thin layer of calcite that may play a stabilizing role and at least reflects the stability of the walls. The flowstone on the central blockfield does not present any fracture. There are no recent rock fragments on it. The enlargement of the room, by collapses, seems currently stopped.

2.8 Coufin Cave (Choranche, Isère, France)

2.8.1 Description

See Figs. 2.54, 2.55, 2.56, 2.57, 2.58, 2.59.

The Coufin Cave was explored in September 1897 by the French caver O. Decombaz. It is currently developed for tourism under the name of Choranche Caves (coordinates: 45° 04′ 29″ N; 5° 23′ 53″ E). From Villard-de-Lans, drive on the D531 to Pont-en-Royans. Two kilometers before the village of Choranche, on the right, a road leads to the show cave. From the car park, a tourist path leads to the entrance of the cave, at the foot of the Urgonian cliffs.

It is a room 55 m long and 45 m wide, with a ceiling reaching 15 m high at its highest point. It is oval in shape and receives, east and south, two underground rivers flowing through two galleries 5 to 10 m wide. The floor of the room is relatively flat; a lake occupies most of it. A trail goes around it and gives access to the two galleries. The ceiling is corbelled and emphasizes the limestone dip with a slab 20 m in span on the side of which a collapse occurred above the lake (see section A–B). The Coufin Chamber belongs to the downstream part of the Coufin-Chevaline Cave network, which extends for 15 km under the Presles Plateau (Fig. 2.56) (Caillaut et al. 1997). The cave is currently used for karst studies (Delannoy et al 2009).

2.8.2 Geology and Hydrogeology

The Coufin Cave is located in the Bourne Valley, 590 m.a.s. l., at the foot of the Urgonian cliff of Choranche Cirque. At the entrance, the room is only at 3 m from the edge of the cliff. This position is similar to that of the Cathedral chamber in the Chevaline Cave, which is also close to the slope (Fig. 2.56).

The cavity is dug in a thick carbonate formation. It is marly at the base and becomes more calcareous upward to finally form the high Urgonian cliff visible in the landscape. The series is cut by large extensional faults bearing N 0° and N45°. These directions are observed in the Coufin-Chevaline Cave network. The room is dug at the base of the Urgonian cliff in the Barremian red sandy limestone (see Bournillon Porch Sect. 2.4; Fig. 2.22). The rock is stratified, with well visible beds 30 to 50 cm thick. The series has a slight dip of 5° to the north.

An N80° joint crosses the chamber in its central part without playing a role in the morphology of the ceiling. Fractures with a similar direction (N100°) are observed in the north gallery, as well as in the south gallery where a cluster of N90° joints is present. These two galleries have a general N5° orientation which is parallel to that of the main fault that crosses the Coulmes Forest.

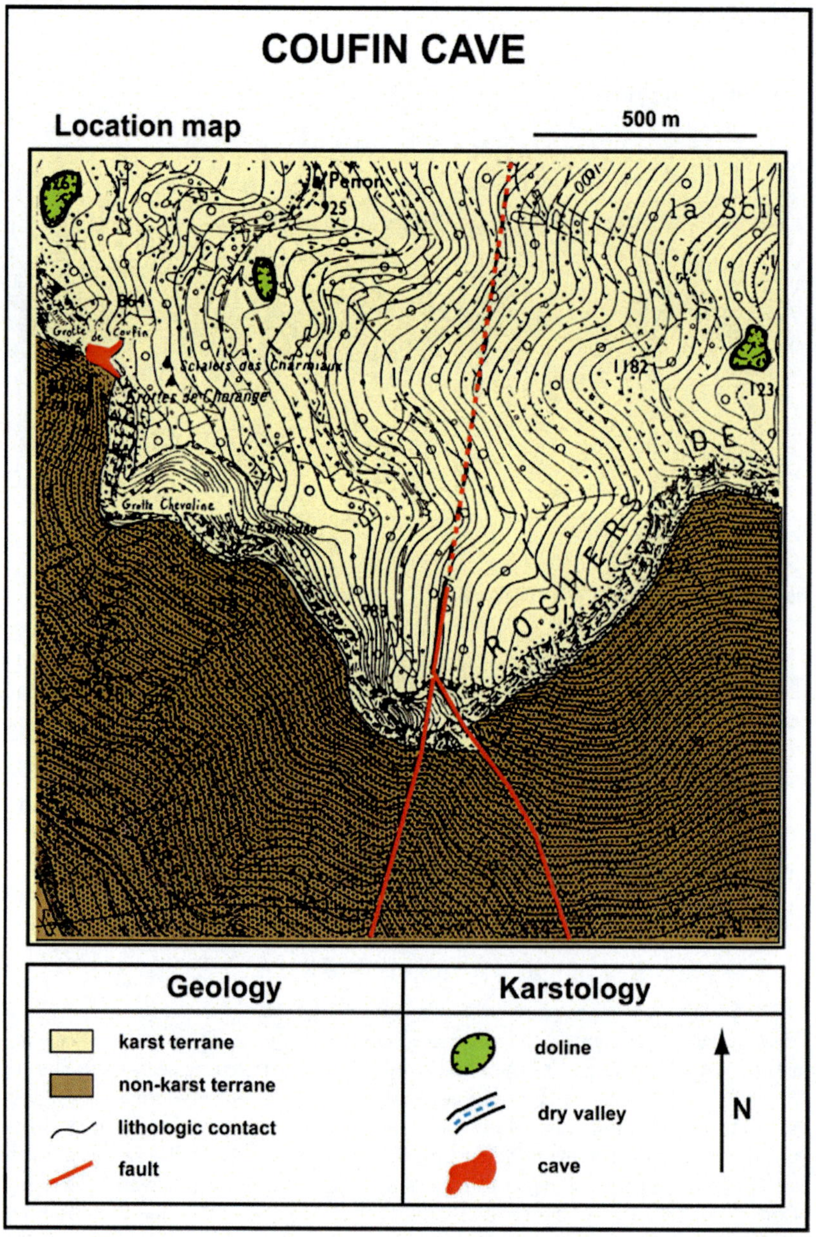

Fig. 2.54 Geomorphological map of Coufin Cave

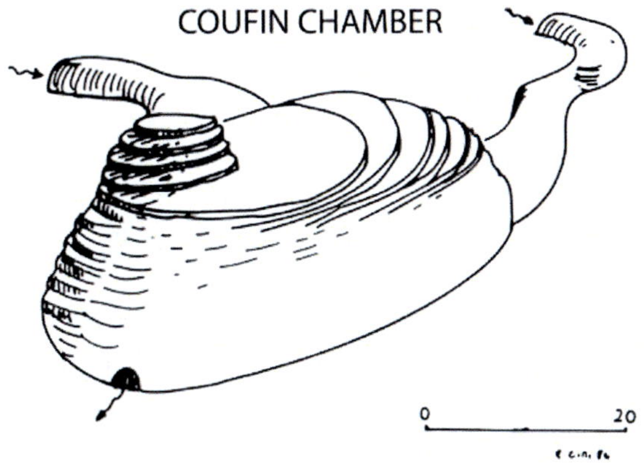

Fig. 2.55 3D sketch of Coufin Chamber

Apart from two blockfield areas, north and east of the chamber, the floor is covered with clay that forms the underground rivers' banks.. Both blockfields are calcified and seem old. Except at the level of the collapse bell, above the lake, the ceiling is decorated by soda straws, exceptional by their density. Their age is difficult to establish but can be considered relatively recent.

The chamber is now the confluence of two underground rivers, one draining the Coufin Cave and the other one that of Chevaline. Both networks join upstream and constitute the drains of the groundwater of the Presles Plateau (Figs. 2.56, 2.58). The impervious base level is formed by the Hauterivian marly-limestone, at the top of which are the springs of the other karst networks of that area (see Sect. 2.4, Bournillon Porch).

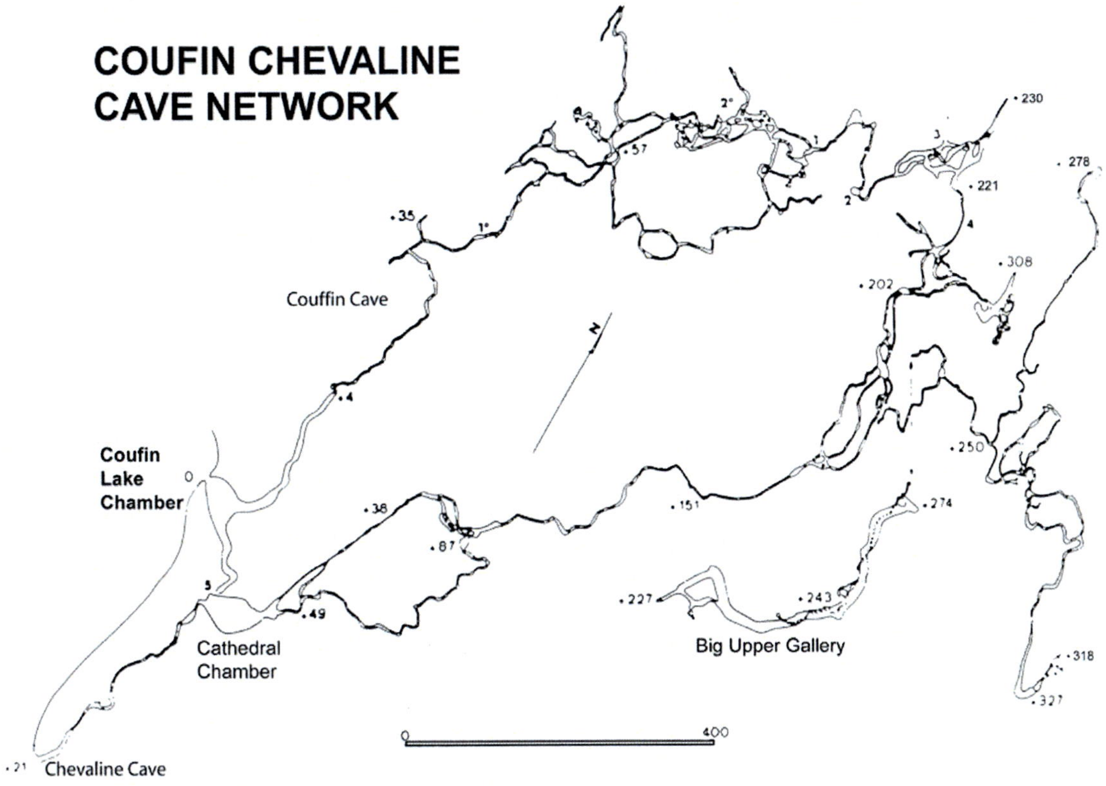

Fig. 2.56 Map of Coufin-Chevaline Cave network

Fig. 2.57 Entrance lake of Coufin Chamber (photo Cla and Clem)

2.8.3 Genesis and Stability

The room is developed by successive collapses of the ceiling, with a clearing off by the rivers. Digging seems to have been done in a tensile strength zone due to the proximity of the slope. The Cathedral Chamber in the nearby Chevaline Cave has a similar position. The galleries of this cave network are parallel to the cliff and also indicate the presence of tensile cracks close to the slope. However, the position of the cliff when the chamber was dug remains to be defined. The collapse bell located above the lake shows flaking phenomena that give the ceiling an appearance of being unstable. The other parts of the chamber look stabilized as evidenced by the absence of blocks on the clay deposits and the presence of soda straws on the ceiling.

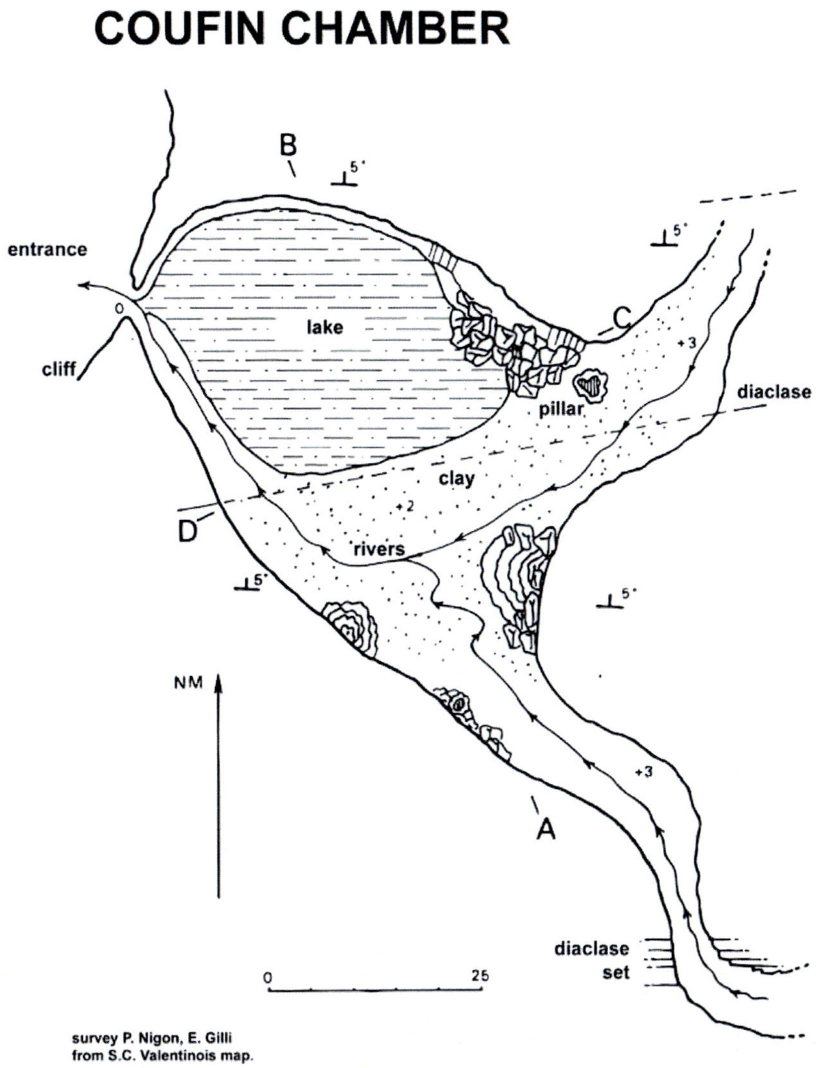

Fig. 2.58 Map of Coufin Chamber

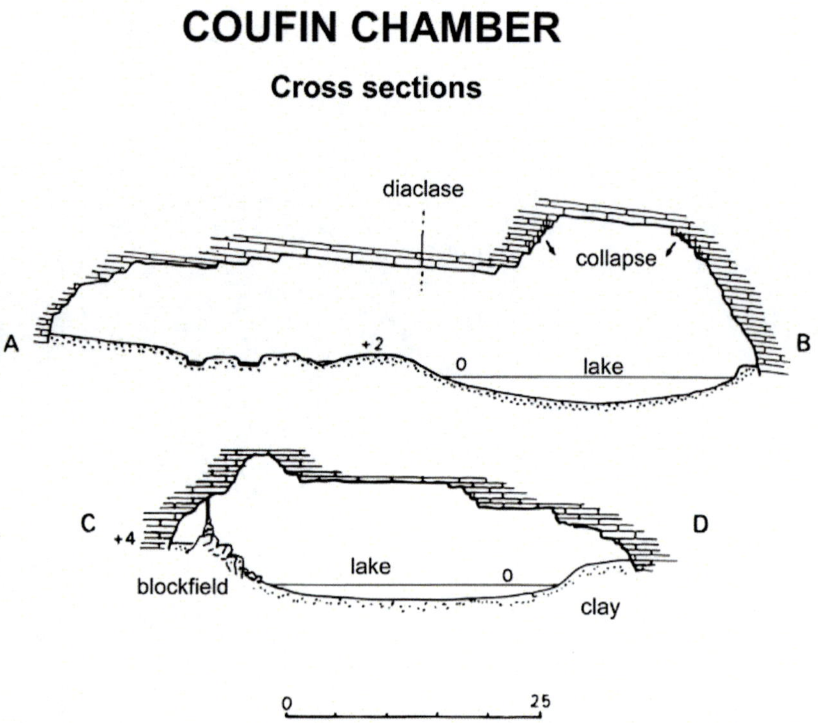

Fig. 2.59 Cross sections of Coufin Chamber

2.9 Fouillac Shaft (Saint-Jean-de-Bruèges, Hérault, France)

2.9.1 Description

See Figs. 2.60, 2.61, 2.62, 2.63, 2.64, 2.65.

From Montpellier drive on the N 109 for 30 kms to Gignac, then take the D9 to the northwest, for 22 km, toward La-Vacquerie and Saint-Martin-de-Castries, until the hamlet of Trivalle. Then drive to the north on the D130, about 10 km to the hamlet of Le-Coulet. A track continues toward the northeast and goes down in a valley after Les-Natges; then it goes up on the northern flank of the Séranne Mountain and passes near the ruins of La-Sauvié. It continues to climb toward the south, until a pass at 689 m altitude, crossed by a high-voltage line clearly visible in the landscape. A well visible path descends to the west. The entrance to the chasm is 100 m north of the trail, 500 m from the pass. The entrance is not very visible. It is located in the center of a bunch of trees (Approximate coordinates: 43° 49' 58.26" N; 3° 35' 2.12" E).

An 18-m vertical pitch leads to the top of a scree cone located at the highest point of a 110-m-long and 80-m-wide chamber. It has an elongated oval shape with a NW-SE axis. The floor and the ceiling are inclined 30° to the east. The floor is covered with a scree covered with flowstone. The lowest part is occupied by a large rimstone pool above which the ceiling rises and forms a large chimney that gives access to upper galleries. The chamber is well decorated, with many stalactites and stalagmites of which some are of large size.

2.9.2 Geology and Hydrogeology

The Fouillac Shaft is located on the northern flank of the Séranne Mountain, which is a complex anticline that forms the southern border of the Causse-du-Larzac Plateau. To the northwest, the La-Vis Canyon separates the Séranne Mountain from the Causse-de-Blandas Plateau. To the southeast, it overlooks the Causse-de-la-Selle Plateau and forms an important escarpment that corresponds to the passage of the Cévennes Fault which is the main tectonic accident of this area.

The entrance to the cave is located at about 615 m above sea level on a small plateau on the left bank of a thalweg which leads to the northwest to the Combe des Natges, a dry tributary valley of La-Vis River. The thickness of limestone above the ceiling is approximately 15 m at the entrance and 45 m above the lowest point of the chamber.

The cave is in an area where the structure is roughly tabular with a slight dipping of the limestone beds toward

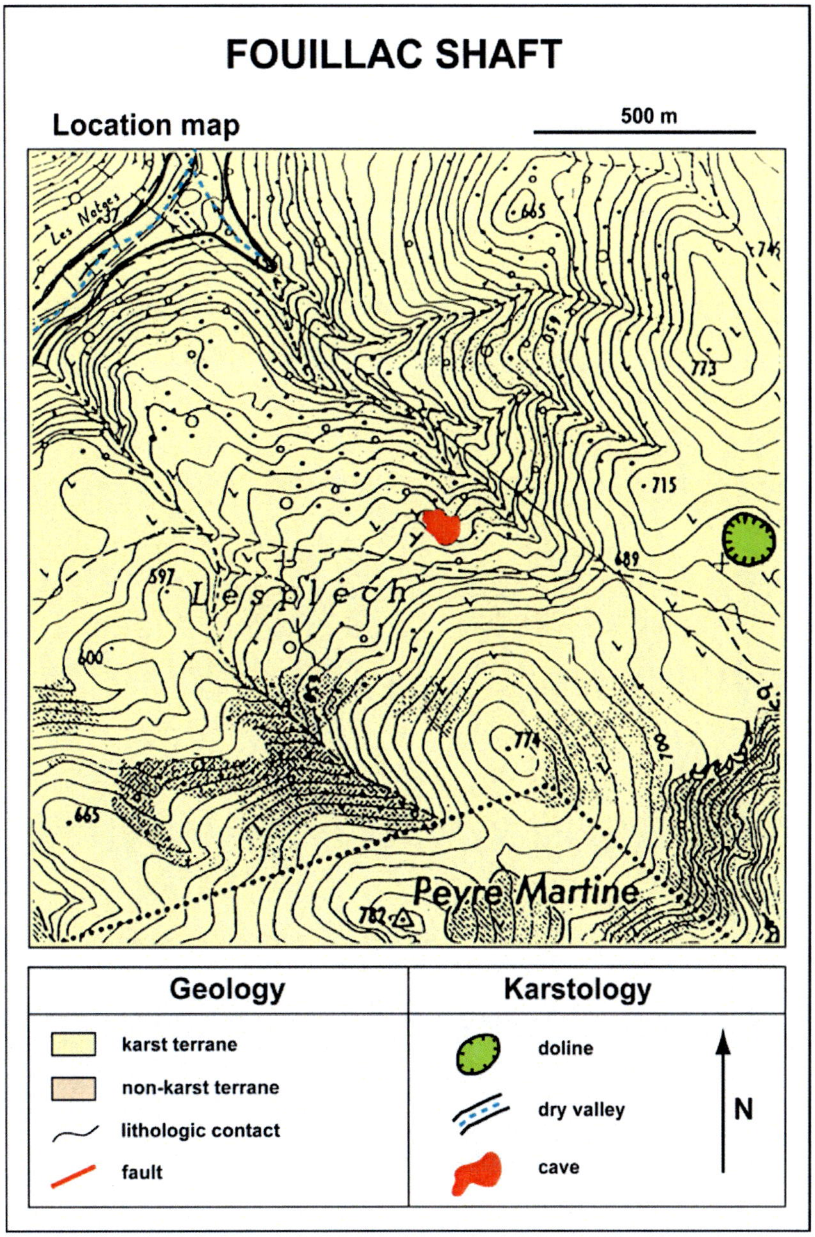

Fig. 2.60 Geomorphological map of Fouillace Shaft

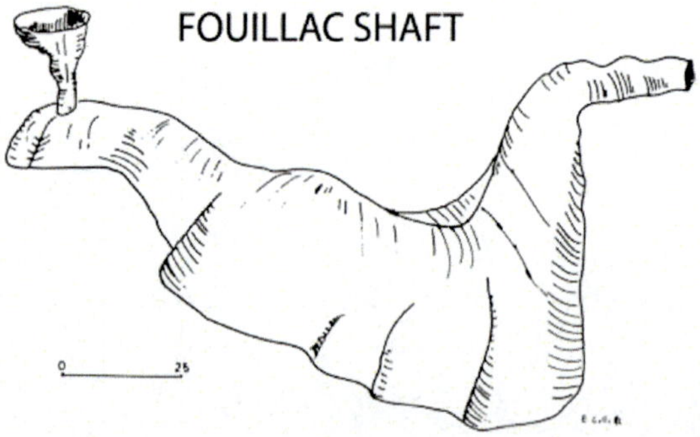

Fig. 2.61 3D sketch of Fouillac Shaft

the northwest in the direction of La-Vis Valley. Two kilometers to the southeast, there is a large cluster of N45° faults (the Cévennes Fault). A parallel fault intersects the series at the Combe des Natges Valley. Between both faults, the Séranne Mountain is compartmentalized by N90° fractures conjugated with the previously described ones.

The Séranne Mountain is a thick calcareo-dolomitic series dated from Aalenian to Portlandian. The aquiclude is formed by Toarcian marls and clays. The Fouillac Shaft is located at the top of the limestone series in Portlandian reef limestone. The rock is very massive; no stratification is detectable inside the cave. The ceiling and the walls are very fractured. The strikes are varied.

Fig. 2.62 Scree cone at the bottom of the entrance pitch of Fouillac Shaft

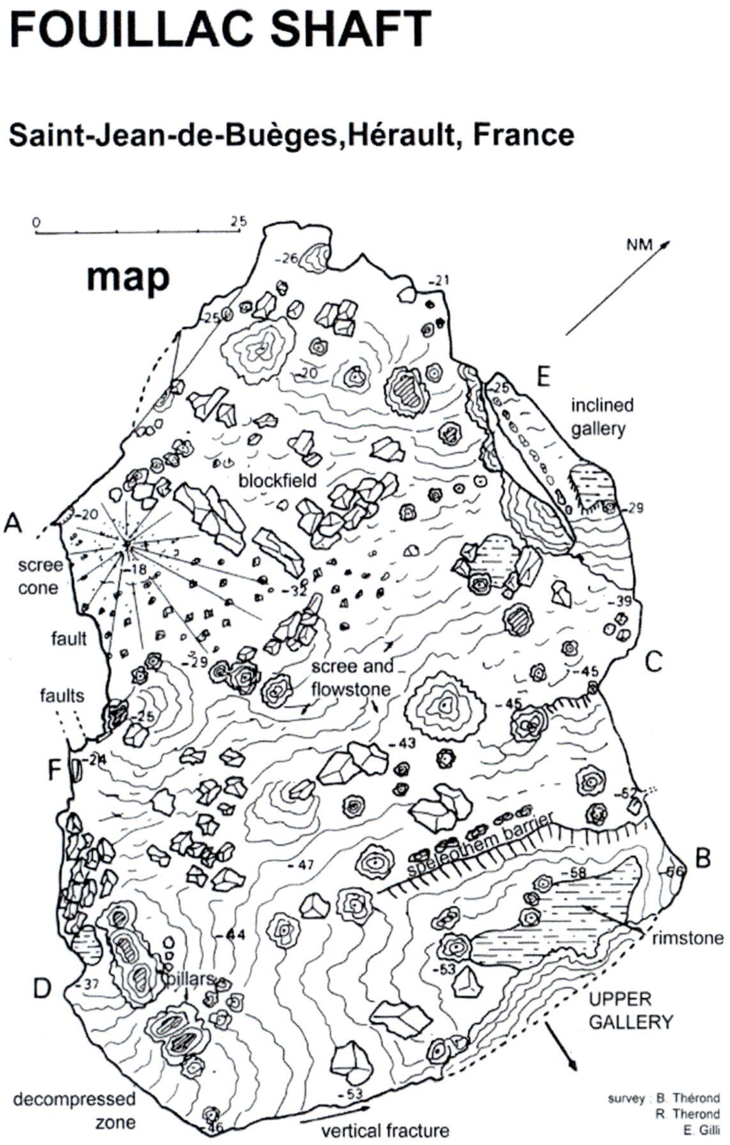

Fig. 2.63 Map of Fouillac Shaft

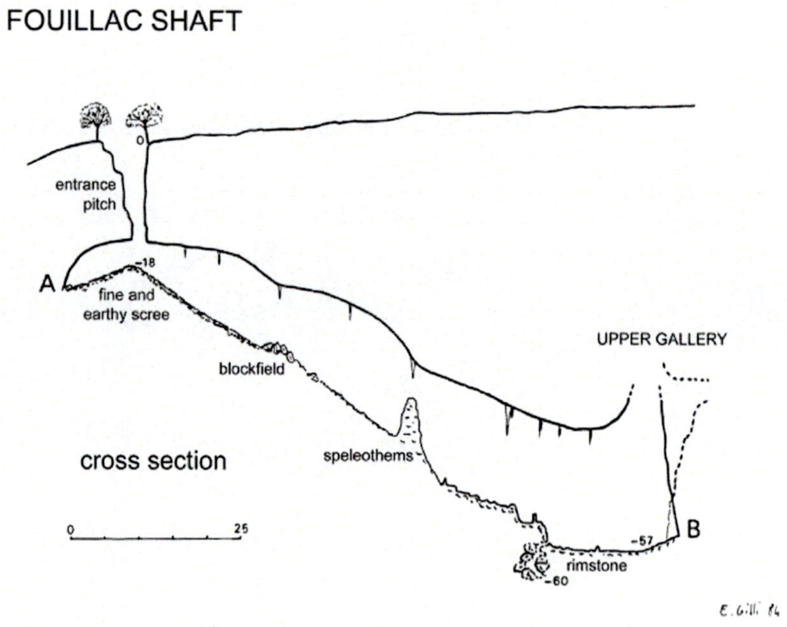

Fig. 2.64 Cross section A–B of Fouillac Shaft

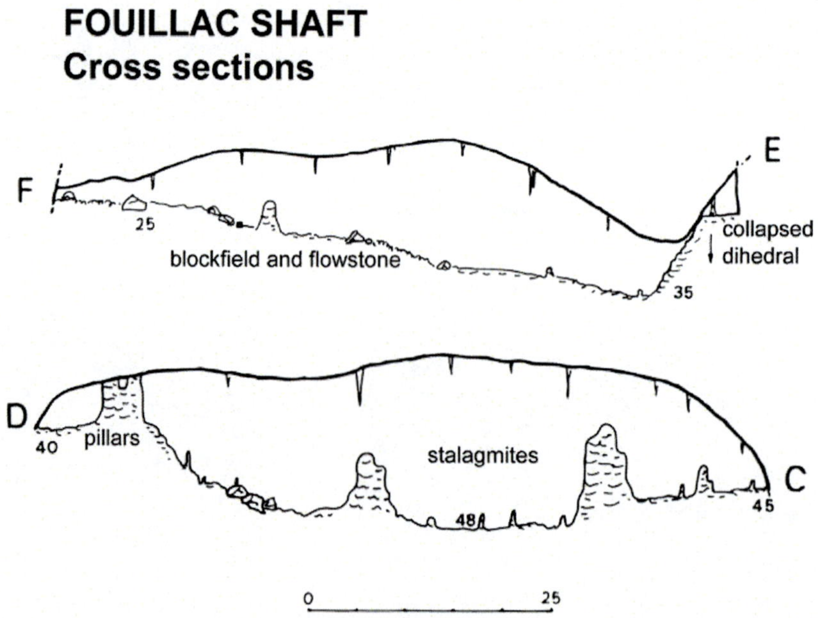

Fig. 2.65 Cross sections C–D and E–F of Fouillac Shaft

The access pitch was created by the collapse of a zone weakened by the presence of a tight network of N30° vertical fractures. The end part of the Fouillac Shaft is formed by a vertical wall presumably aligned on an N8° fault. The southeast wall is crumbled and perhaps corresponds to the wall of an N90° fault. The base of the entrance pitch is occupied by a scree whose relatively fine elements (diameter: 0.1 to 0.2 m) come mainly from the alteration of the pitch walls. This scree is deposited on a large blockfield covered with speleothems and sometimes completely masked by the flowstone. The speleothems are numerous on the ceiling and on the floor. In the bottom part of the chamber, some stalagmites reach a few meters in diameter at their base.

In the deepest part of the chamber in the northeastern corner between the rimstones and the stalagmites, it is possible to get in between the blocks for a few meters without being able to reach the bedrock (see A–B cross section). Flowstone is currently forming. A sheep skull deposited a few years ago at the top of a stalagmite is now completely covered with calcite several millimeters thick. In the end part of the cave, a trickle of water feeds the rimstone pools before sinking between the blocks.

2.9.3 Genesis and Stability

It is a collapse chamber that may result from the emptying of a dolomitic sand pocket located at depth. The flowstone that covers the limestone blockfield is not fractured which proves the stability of the floor. The ceiling is also covered with cave formations, and no recently collapsed rocks are present on the floor (except for the entrance pitch area). Some decompression signs are visible on the south wall of the chamber, mainly in the bottom part of the cave; but the calcite deposit is not fractured which indicates that the decompression phenomenon is currently stopped. As a whole, the chamber seems stabilized with a marked tendency for speleothems deposit.

2.10 Kocain (Huge Cave) (Ahırtaş, Antalya, Turkey)

2.10.1 Description

See Figs. 2.66, 2.67, 2.68, 2.69, 2.70, 2.71, 2.72, 2.73, 2.74, 2.75.

This chamber, the largest one in Turkey, has been known since prehistoric times, as evidenced by the numerous Neolithic pottery shards on the ground. It was occupied in ancient times and contains a cistern and remnants of a temple and Late Roman inscriptions (Harper 2017). Despite its size, it was little known in the region until recently when an access road was built. From Antalya drive on the D650 for 50 km toward Burdur until Selimiye, then turn on the right to Ahirtas village. From there, a track leads to the cave (Fig. 2.66) (coordinates: 37° 13' 57.34" N; 30° 42' 39.73" E).

The hall is rectangular. It is 250 m long and 150 m wide. The access is by a 200-m-long and 70-m-wide gallery. This part, whose entrance is a vast porch, has an earthy soil. It goes down regularly about 20 m lower in altitude. At 150 m from the entrance, on the right, Greek cisterns collect a trickle of water pouring from the top of a flowstone. The gallery ends on a huge blockfield that can be skirted by the right. After about 50 m on a flat and loamy ground, there is a pass between two large stalagmites visible from the entrance. Then a climb on the blockfield makes it possible to reach the highest point of the chamber occupied by two huge stalagmitic pillars, 30 m high. From there, the scree goes down to the southwest for a few tenth of meters. There is no known further passage.

The western wall of the cave is a plane inclined at 30°; the eastern wall is more upright. The ceiling is arched in the access gallery zone (Fig. 2.72) and has an M shape at the highest point of the chamber (Fig. 2.74). At the bottom of the hall, the ceiling is quite regular. Fractures that affect the limestone massif, parallel to the long axis of the cave, are at the origin of the shape and the inclination of the western wall.

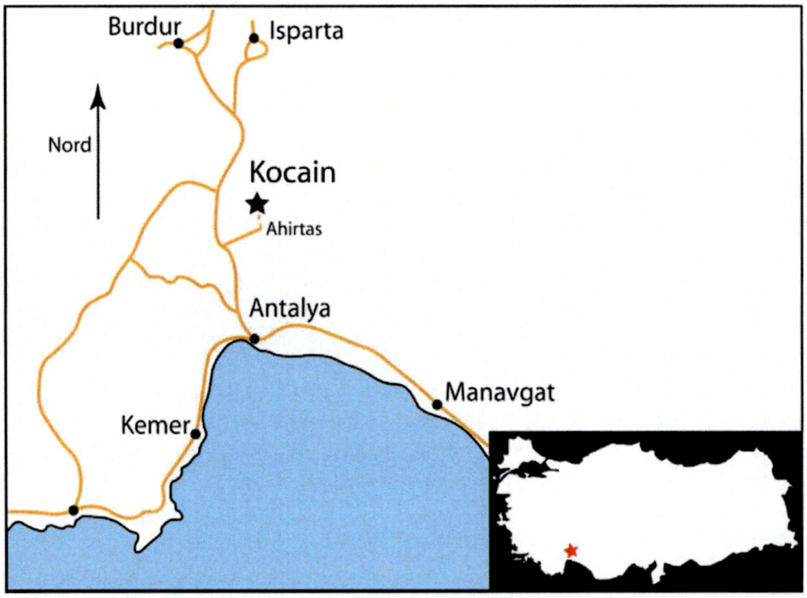

Fig. 2.66 Location sketch of Kocain Cave

Fig. 2.67 Gigantic entrance of Kocain Cave

Fig. 2.68 Main gallery of Kocain. Antique cisterns are visible on the right

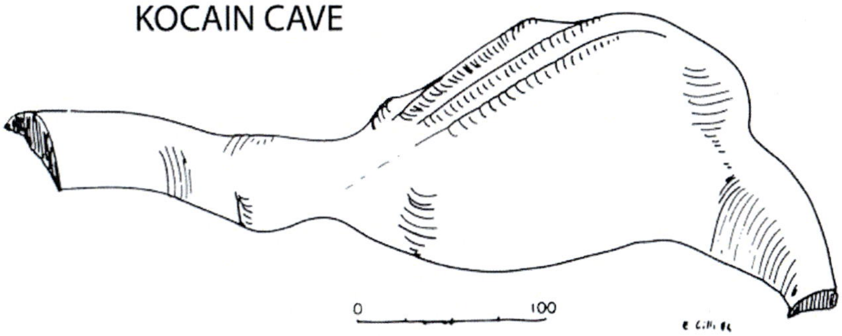

Fig. 2.69 3D sketch of Kocain Cave

Fig. 2.70 Main speleothems of Kocain Cave (photo A. E. Keskin)

2.10.2 Geology and Hydrogeology

Kocain Cave is on the northeast slope of a mountain and plunges perpendicular to the interior of the massif. On the surface, the limestone is intensely eroded which makes walking off the trails very difficult. The thickness of the limestone above the ceiling is 30 m at the entrance of the cave and increases very rapidly to reach around 100 m above the chamber. The rock is a beige microfractured massive limestone. The bedding is not visible into the cave but can be observed outside. The strike is N0°; the dip is 30° to the east. The cave is dug in very tectonized folded series. A set of large fractures oriented N45° and inclined toward the east emphasize the main axis of the cave. These accidents constitute a tight network clearly visible on the ceiling above the large stalagmites (Figs. 2.73, 2.74).

There are two areas:

- The entrance gallery whose earthy floor contains small blocks.
- The actual chamber occupied by a blockfield of several meters sized elements. These blocks are covered by flowstone that sometimes masks them completely (southern end of the room).

The walls and the ceiling are covered with flowstone and speleothems, and this makes geological observation difficult. The large size of the stalagmitic pillars at the back of the room (10 m in diameter and more than 30 m high) testifies to the old age of the speleothems. Apart from a few trickles of water, there is no water circulation. However, the morphology of the access gallery shows that it is an old karst drain.

2.10.3 Genesis and Stability

The Kocain Chamber is a local widening of a giant gallery, provoked by the collapse of the ceiling, related to the large fractures that are visible in the ceiling (Fig. 2.68). The Kocain Chamber is perfectly stable as evidenced by the important deposit of flowstone on its floor.

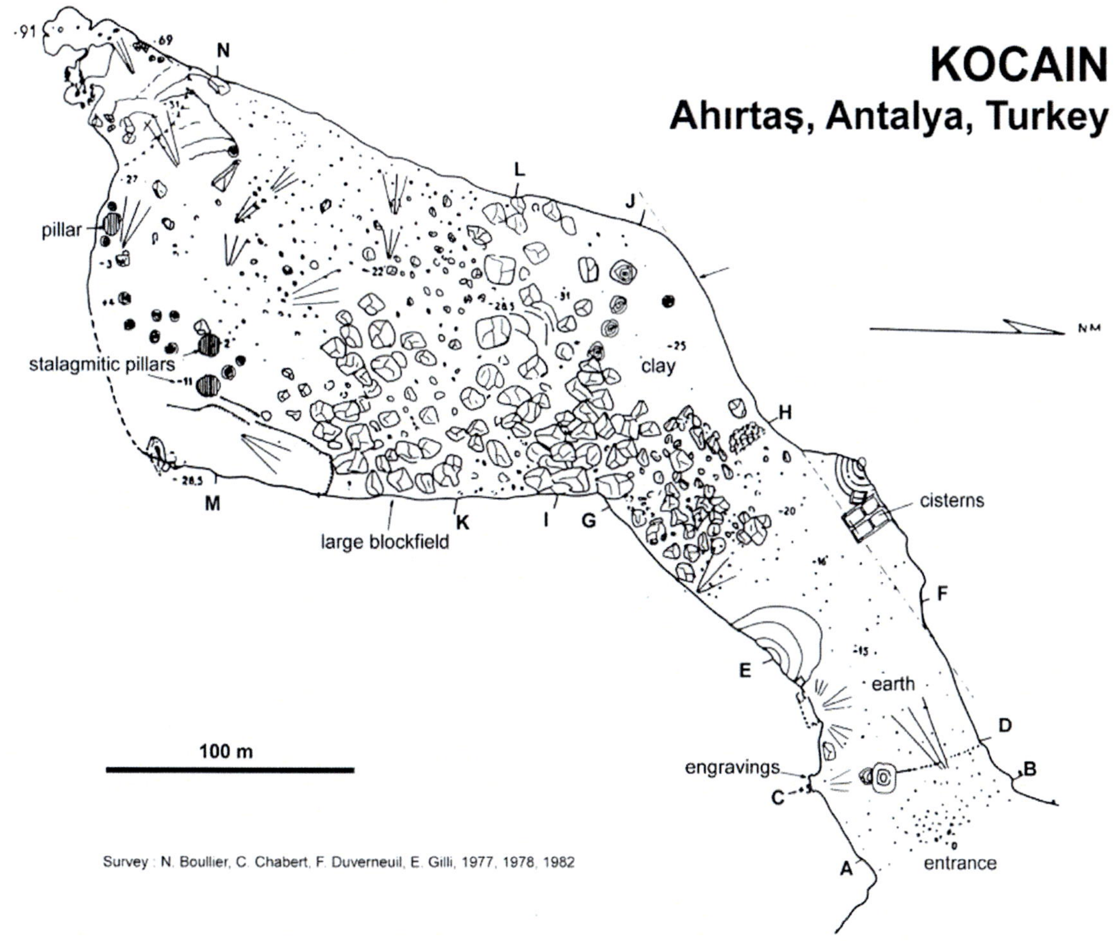

Fig. 2.71 Map of Kocain Cave

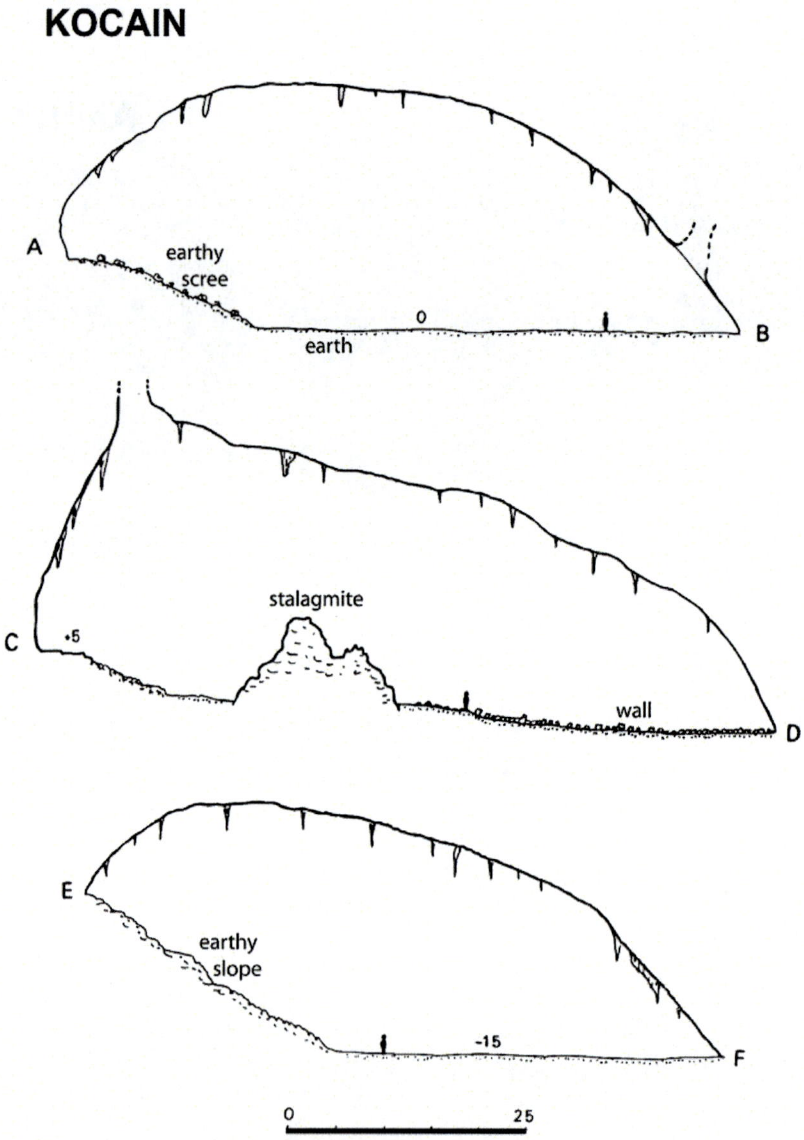

Fig. 2.72 Cross sections A–B. C–D and E–F of Kocain Cave

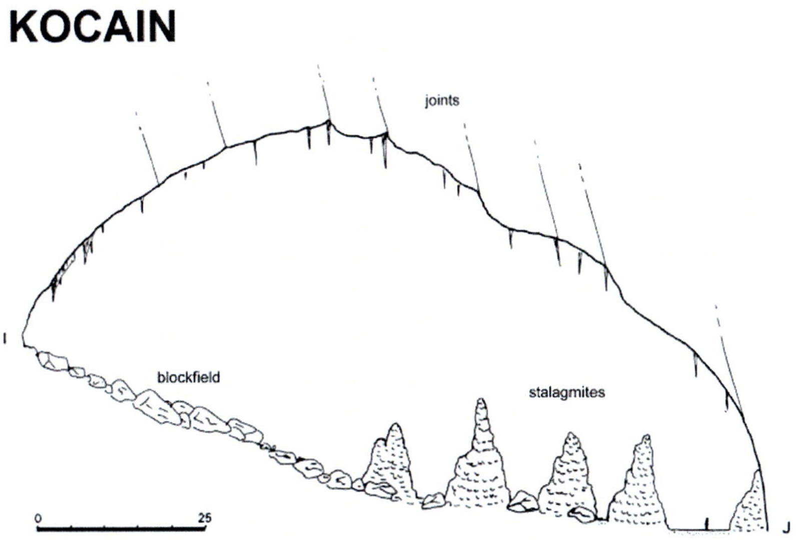

Fig. 2.73 Cross section I-J of Kocain Cave

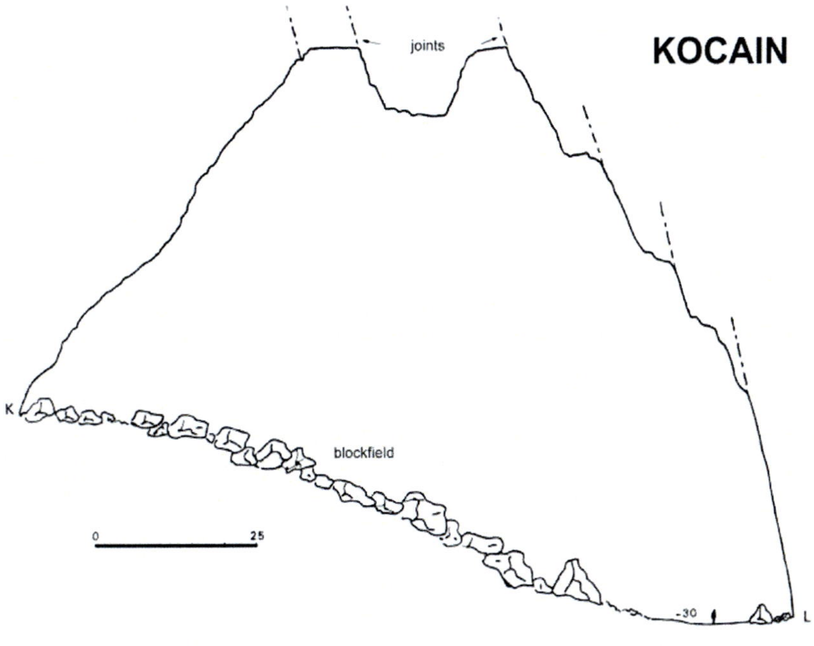

Fig. 2.74 Cross section K-L of Kocain Cave

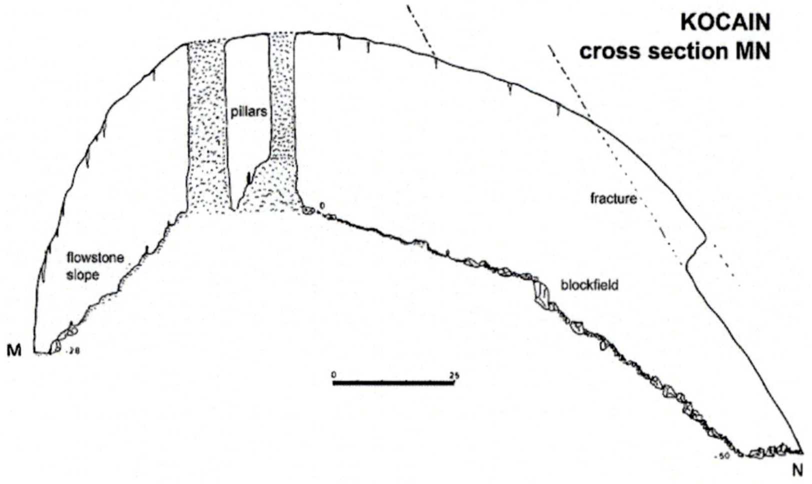

Fig. 2.75 Cross section M-N of Kocain Cave

2.11 Mas D'Azil Cave (Mas D'Azil, Ariège, France)

2.11.1 Description

See Figs. 2.76, 2.77, 2.78, 2.79, 2.80, 2.81, 2.82, 2.83.

The cave is located 1 km from Mas d'Azil village. It has been known since prehistoric times. It gave birth to the Azilian prehistoric civilization that produced enigmatic painted or carved pebbles (Azéma and Brazier 2016). At various times, it was used as a shelter. During the French Wars of Religion (1562–98), it welcomed the Protestant population of Mas d'Azil village. It was also used as a shelter for a plane engine factory during WW2 (Pétris 2017). Nowadays, it is developed for tourism and serves as a natural tunnel to the D119 road that crosses the cave toward St. Girons (coordinates: 43° 4' 4.75" N; 1° 21' 16.17" E).

The cave of Mas d'Azil is a 450-m-long S-shaped gallery. It is crossed by the Arize River. It can be divided into two parts. From upstream to downstream,

- The entrance zone: In this area, the gallery is 60 to 100 m wide, with ceiling heights of 40 m and important slab spans.
- The downstream zone: The dimensions are lower with width from 30 to 50 m and ceiling height less than 20 m.

Only the upstream zone is therefore part of this study. In this area, the cave opens with a monumental porch of square section, whose dimensions are 45 × 45 m. The gallery widens immediately after. The Arize River flows in its center. The left bank is occupied by an earthy slope 15 m high, which took place in prehistoric excavations. The right bank is formed by a smaller embankment in which the road is partly dug. Both slopes are lowered downstream and give way to banks of pebbles and big boulders. After 250 m the gallery narrows, the ceiling lowers and the downstream part is reached.

On the right bank, about 15 m above the river, at the level of the road, a car park has been built in an enlargement of the gallery and provides access to a vast labyrinth where a museum of prehistory is installed. The cave is crossed by the Arize River. The discharge is around 500 L/s. It varies according to the seasons and can be null during the dry period.

2.11.2 Geology and Hydrogeology

The entrance to the cave is 320 m.a.s.l: the exit is 303 m. The river crosses a rocky outcrop, which is only 300 m wide in the cave zone. This small mountain, bordered by cliffs, is the western end of a vast syncline. Above the cave, a little to the east, at the Baudet farm, a depression in the landscape

Fig. 2.76 Upstream entrance of Mas d'Azil Cave

Fig. 2.77 Road inside the Mas d'Azil Cave. Note the flat ceiling (Photo NC SA Creative Common)

corresponds to an ancient riverbed of the Arize before its sinks into the underlying limestone (Fig. 2.78). The thickness of limestone above the cave is 60 m. The cave intersects the Pradals syncline of the E.W. axis, with low dipping. A double set of faults of NE-SW and ENE-WSW directions crosses the syncline in the zone where the cave is located.

From bottom to top, lithology is as follows (Fig. 2.81):

- 30 m of clayey nodular limestone that forms a poorly compact unit, which is very sensitive to erosion. It is in this unit that the widest parts of the cave are hollowed;
- 10 m of more compact nodular limestone;
- 10 m of very compact pink limestone in beds 1–2 m thick. They form the ceiling of the cave;
- a very large bench of massive limestone, more than 10 m thick, overcomes the cave. This level is clearly visible above the porch at the upstream entrance.

These formations are dated Thanetian. At the upstream entrance, the strike of the layers is N20°, and the dip is 10° to the west. The downstream exit is located on the other flank of the syncline; the strike is N0° with an 18° dip to the west. The fault network drawn on the 1/50,000° geological map is not visible inside the gallery. In the cave, only large joints can be seen crossing the ceiling without apparently affecting its stability (Fig. 2.80).

Cave fillings are of two types:

The terraces of the entrance: They are about 15 m thick with some large blocks embedded in a large earthy mass. They probably date from the last ice phase during which a decrease in the flow of the Arize resulted in the deposition of fine elements. A large part of these terraces was later eroded, no doubt during the melting of glaciers.

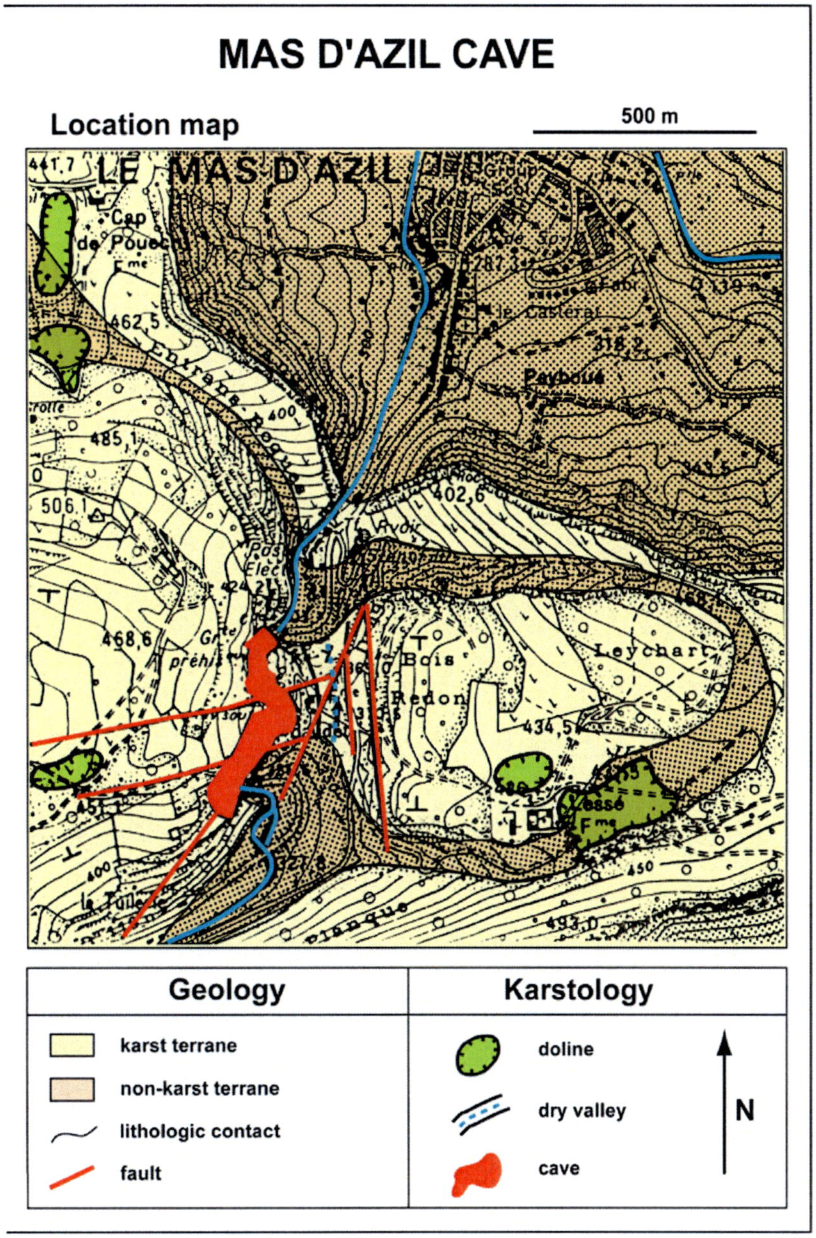

Fig. 2.78 Geomorphological map of Mas d'Azil Cave

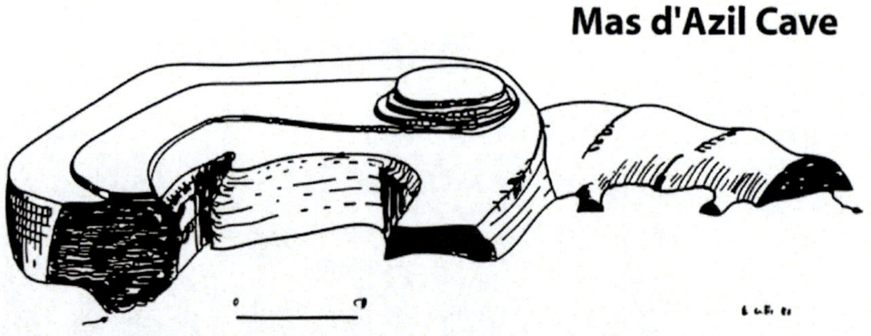

Fig. 2.79 3D sketch of Mas d'Azil Cave

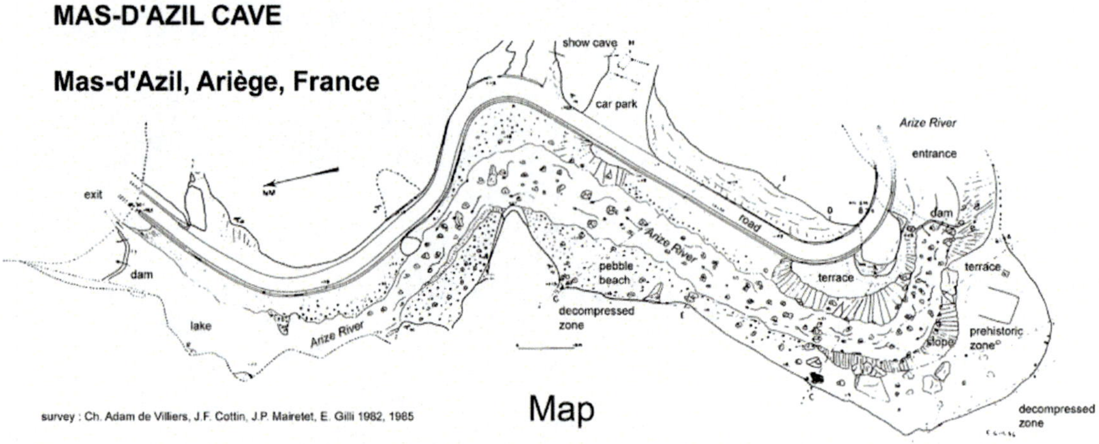

Fig. 2.80 Map of Mas d'Azil Cave

<u>The riverbed:</u> It is cluttered with boulders and pebbles. The blocks are witnesses of the widening of the cavity by collapse. The pebbles result in the action of the river on the blocks. The amount of blocks decreases from upstream to downstream of the gallery. The cavity is devoid of speleothems.

2.11.3 Genesis and Stability

The Mas d'Azil was formed by the swallowing in the limestone of the Arize River which previously flowed above the syncline in the zone where the Baudet farm is. This underground capture may have been caused by the deepening of the bed of the Arize River which runs on thick marl series downstream of the limestone syncline. The rupture of the river profile would then have been mitigated by the creation of an underground shortcut.

The digging occurred at the level of the nodular clayey limestone. It vertically stopped when reaching the compact limestone beds that now form the ceiling. The important discharge of the Arize Rivers allowed both a mechanical erosion and a rapid clearing of the erosional products. Chipping is observed at the foot of the vault, on the left bank, in the prehistoric terrace area and at the northern end of the cave on the same bank. The rest of the cavity is stable, except for cryoclastic spalling near the entrances. The filling of blocks and pebbles immersed in the river may be undergoing dissolution.

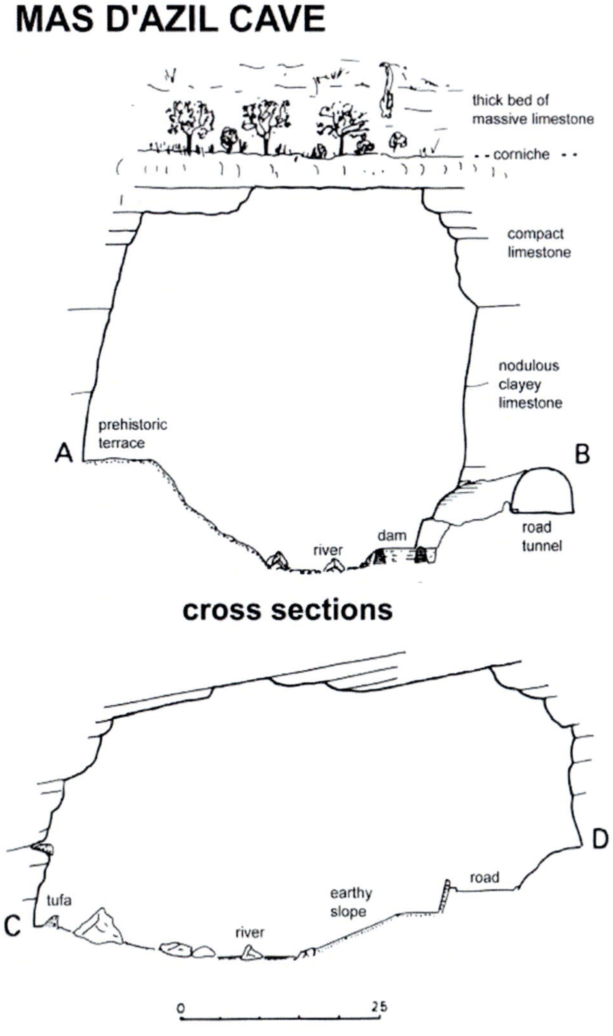

Fig. 2.81 Cross sections A-B and C–D of Mas d'Azil Cave

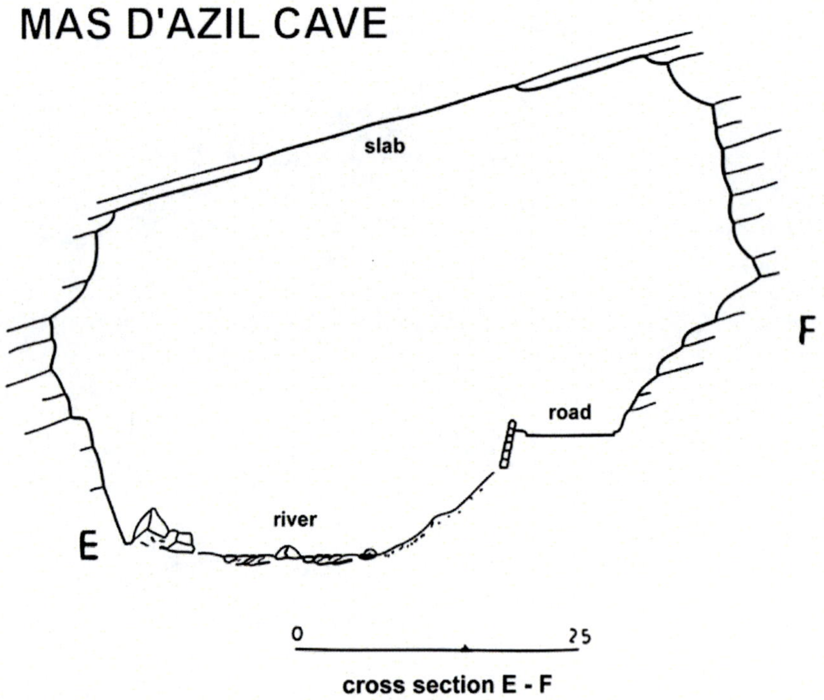

Fig. 2.82 Cross section E–F of Mas d'Azil Cave

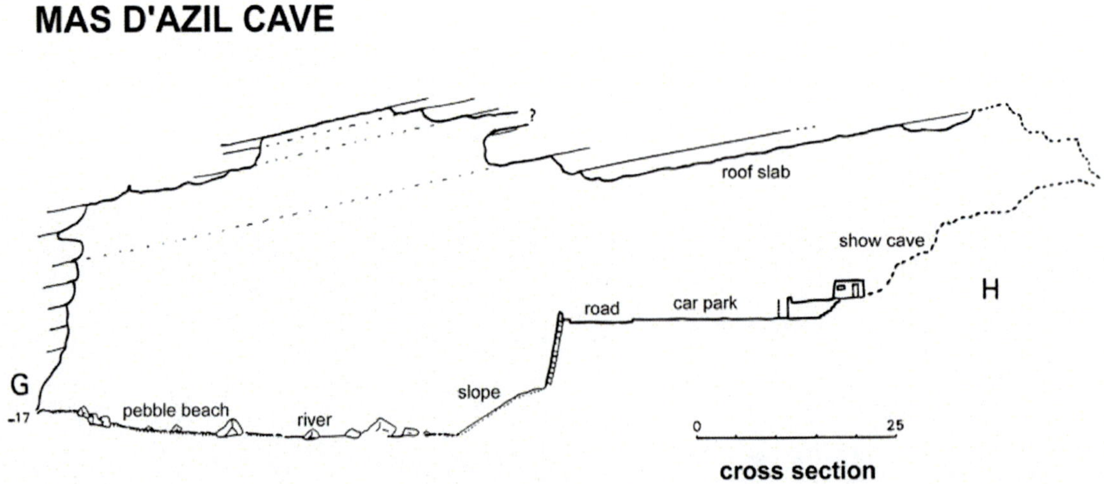

Fig. 2.83 Cross section G-H of Mas d'Azil Cave

2.12 Moras Cave (Rencurel, Isère, France)

2.12.1 Description

See Figs. 2.84, 2.85, 2.86, 2.87, 2.88, 2.89.

This cave, which has probably always been known, is located in the Coulmes State Forest. It is accessed from Villard-de-Lans by the D531 until Balme-de-Rencurel, then by the D35 to the north. After Reppelins hamlet, a trail climbs to the left. After 1300 m, leave the car near a fence that borders a large meadow on the right. The cave is at a 100 m from the track at the bottom of a sinkhole (coordinates: 45° 7' 18.21"N; 5° 27' 57.75" E). It does not require any hardware to reach it.

The Moras Cave is a room 60 m wide and 70 m long. The ceiling is relatively flat at a height of 20 m. The access

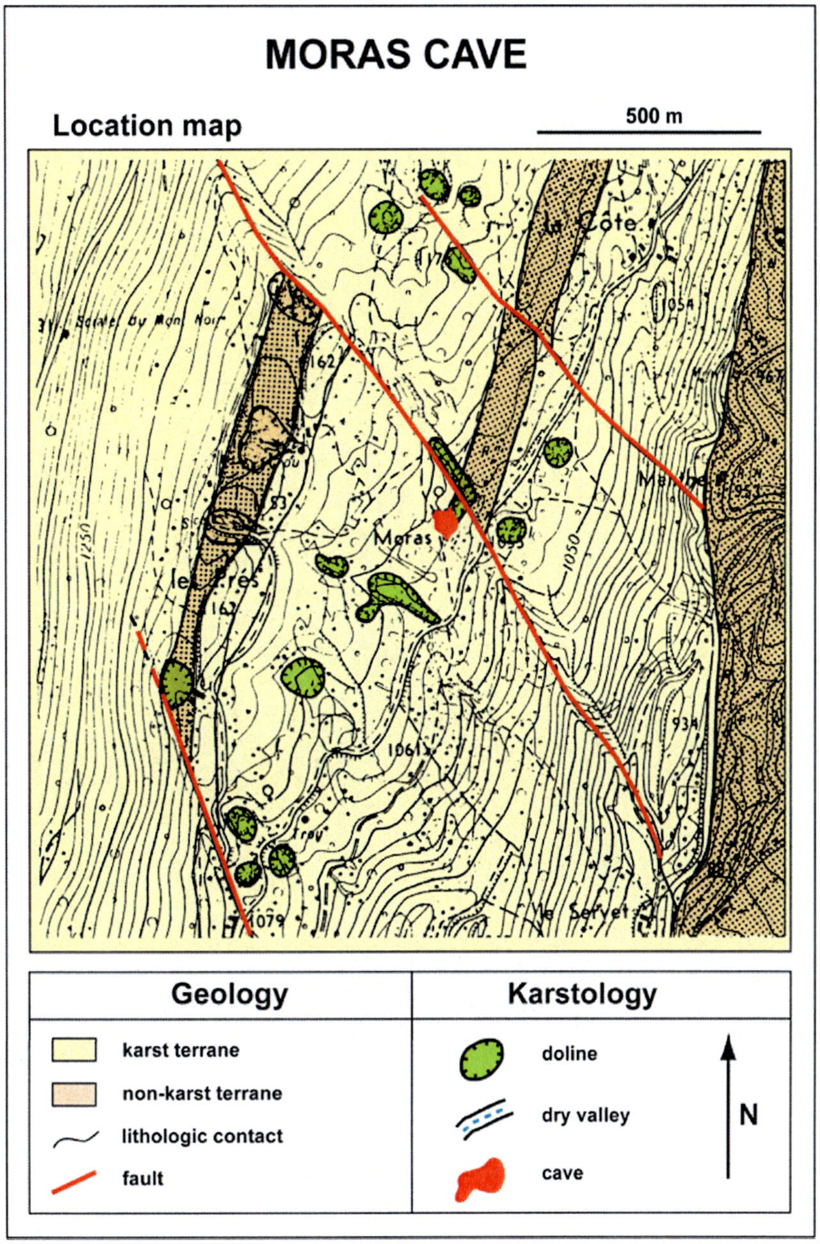

Fig. 2.84 Geomorphological map of Moras Cave

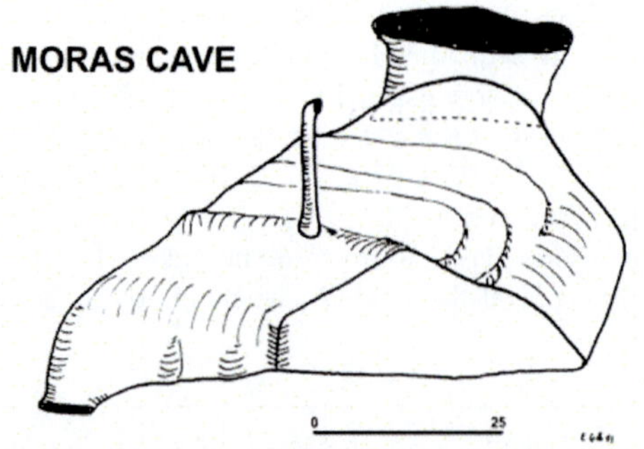

Fig. 2.85 3D sketch of Moras Cave

is by a steep slope until a porch. From there, a scree slope inclined at 20° leads to the bottom of the room where there is a blockfield formed by limestone slabs collapsed from the ceiling. The floor of the room is round-shaped. The south-west side is the wall of an almost vertical fault that extends on about 10 m.

The ceiling is relatively flat in the eastern half of the room. In the western side, it curves to reach the floor. The bedding is well visible. In the northwest corner, in a recess of the wall, the rock exhibits a phenomenon of flaking and detachment of the limestone beds (see cross section H—Fig. 2.88). A trickle of water, which created rimstones, flows from the wall in the back of the room.

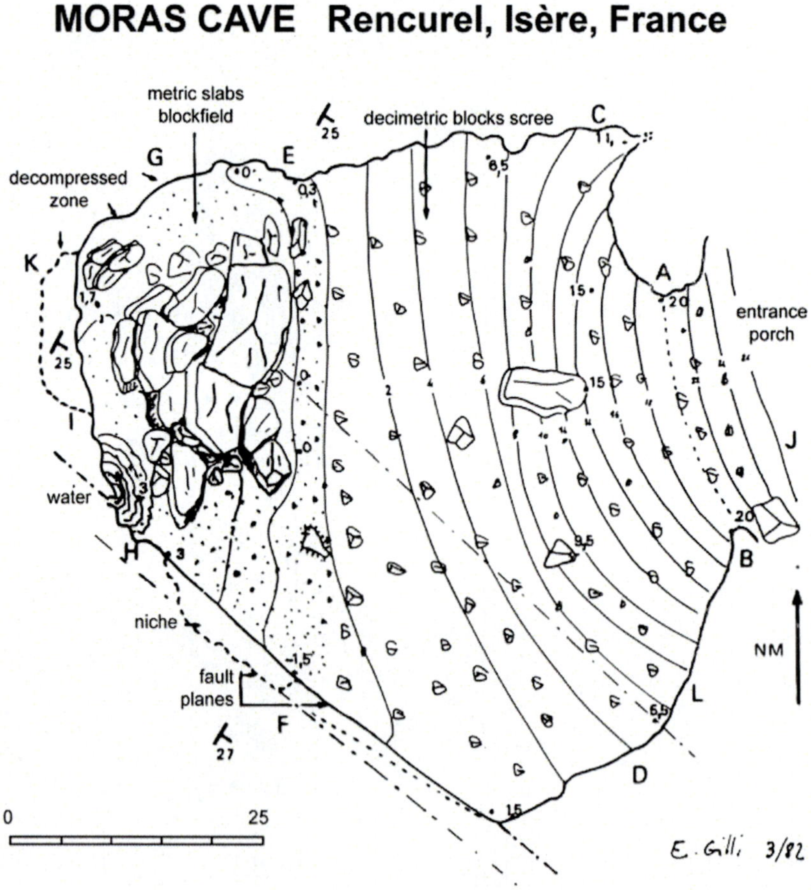

Fig. 2.86 Map of Moras Cave

Fig. 2.87 Bottom part of Moras Cave

2.12.2 Geology

The cave is located at 1075 m.a.s.l. on a very karstified plate marked by numerous sinkholes on the side of the Rencurel Valley. The entrance is located at the bottom of a sinkhole that marks the end of a trough aligned on an N140° fault at the foot of an escarpment of same orientation. The thickness of limestone above the cave is around 30 m.

The cave is located on the edge of the carbonate series of Presles Plateau that dips toward the Rencurel graben. This trench is framed by large N5° faults associated to N140° secondary faults. The entrance of the cave is located close to one of these later faults.

The room is entirely dug in the Senonian. It consists of a series of 5-cm-thick layers of sandy limestone, forming 1-m-thick very compact beds separated by thin marly interbeds. The layers strike is N40° with a 25° dip to the east. A large N140° fault forms the southwestern wall of the cavity. A diaclase of the same orientation as the fault crosses the room in its center. It is highlighted by many stalactites.

There are two distinct parts of cave filling. The bottom of the room is relatively flat. It is occupied by a blockfield of large slabs which sometimes are a few meters long. These slabs come from the ceiling by parting of the limestone at the level of marly interbeds. The phenomenon is observed above a recess in the south corner of the room.

In the entrance area, a thick scree gradually filled the cave. It consists of medium-sized elements with decimetric blocks and a few big ones. This scree is mainly fed by the collapse of the roof but also by the withdrawal of materials from outside. The size and shape of the blocks suggest a cryoclastic origin for part of the scree, mainly in the entrance porch zone.

The ceiling contains some speleothems, especially along the central diaclase. The stalactites are more numerous toward the bottom of the room. The inclination of some of them is related to an airflow. Some small active rimstones are present at the foot of the wall in the bottom of the chamber.

2.12.3 Genesis and Stability

It is a collapse room. The ascending erosion of the ceiling reached the bottom of the sinkhole and created the current entrance access and the huge pile of scree. The original cave did not leave any traces in the studied zone. In the bottom part of the chamber, the relatively developed speleothems show the stability of the ceiling. In the entrance zone, the

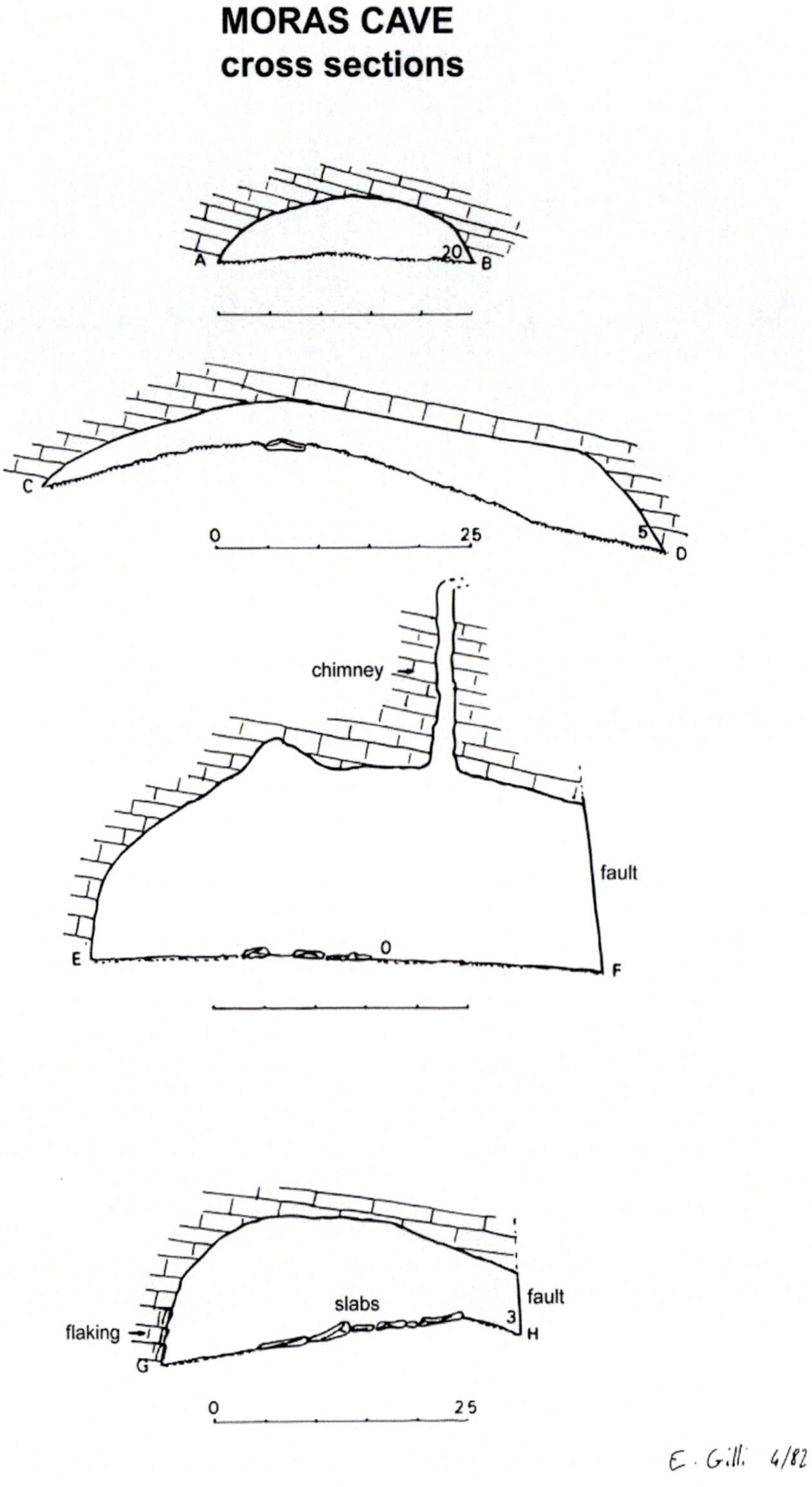

Fig. 2.88 Cross sections A–B, C–D, E–F and G–H of Moras Cave

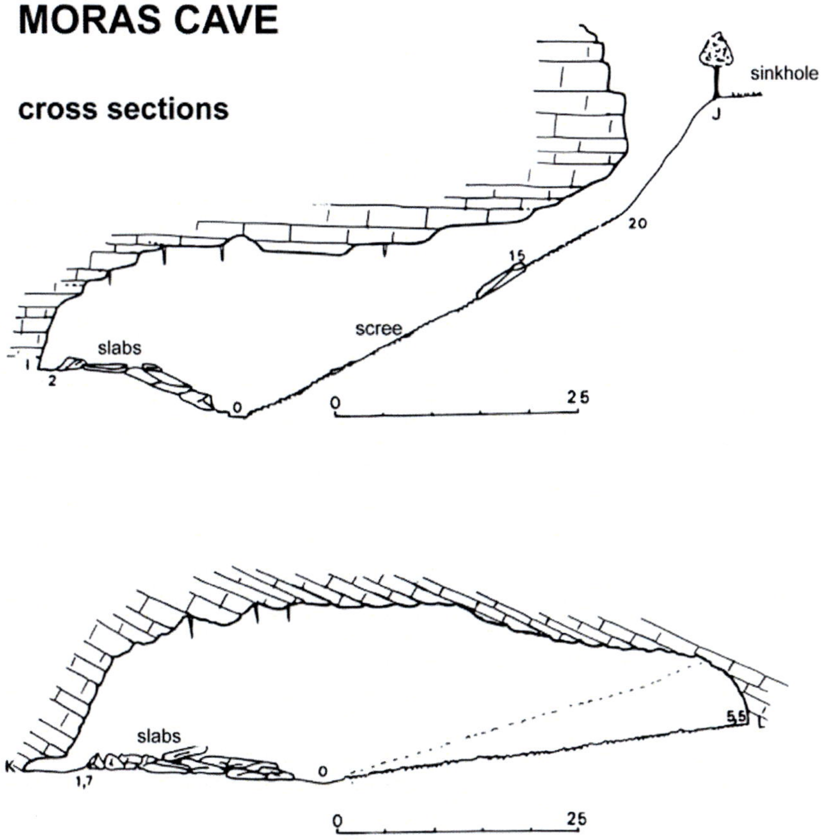

Fig. 2.89 Cross sections I-J and K-L of Moras Cave

cryoclasty causes small flaking or falling of small blocks without any effect on the stability of the whole; however, this does not make possible the growth of stalactites.

2.13 Mort Ru Cave (Saint-Mème, Isère, France)

2.13.1 Description

See Figs. 2.90, 2.91, 2.92, 2.93, 2.94, 2.95, 2.96, 2.97, 2.98.

From Saint-Pierre-d'Entremont, take the D45C road to Saint-Même Cirque until reaching the chalet of Touring Club de France. A few meters before the chalet a path climbs to the limestone cliff. It follows the Mort Ru Valley which begins at the entrance of the cave (approximate coordinates: 45° 23' 53.57" N; 5° 53' 53.89" E).

The cave is part of an important cave network that extends in the Seuil Massif (Loiseleur 1994). A porch a few meters wide gives access to a room 90 m long and 50 m wide. It can be divided into two parts. It is characterized by curiously regular shapes:

- at the entrance, a dome 50 m in diameter;
- toward the bottom, a rising gallery shaped like a nef, with vertical walls and a rounded vault (see Fig. 2.98).

Between these two parts on the east wall, there is a natural rock pillar. The vaults are about 30 m high. The floor is cluttered with scree with several big blocks of metric size. A thalweg cuts the scree in all its length. The top of the room is the junction between an upper network of galleries and a lower one.

2.13.2 Geology and Hydrogeology

The cave opens at 1120 m altitude at the foot of the thick cliff of the Saint-Même Cirque. It is located at the beginning of a thalweg which corresponds to the dry bed of the river that flowed out of the cave.

The Urgonian series is divided into two sequences separated by a more marly level: the Orbitolines layer. The "lower mass" forms in the landscape a plateau surmounted by the "superior mass". It is under this shelf that the room is

Fig. 2.90 Mort Ru Chamber

Fig. 2.91 Natural pillar in Mort Ru Cave

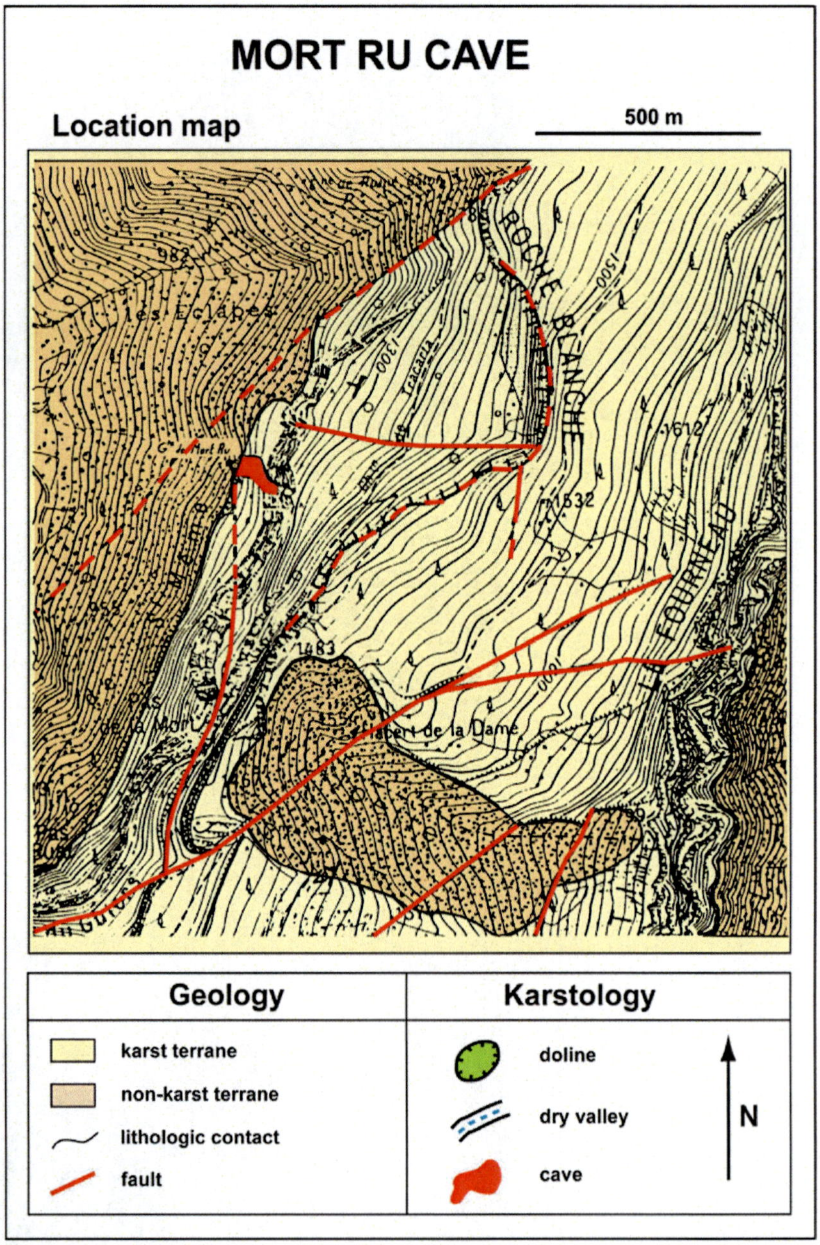

Fig. 2.92 Geomorphological map of Mort Ru Cave

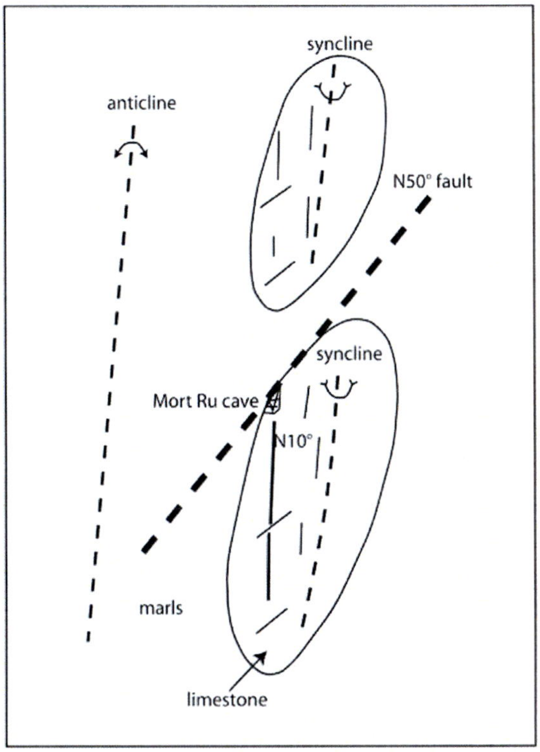

Fig. 2.93 Structural scheme of mort Ru Cave

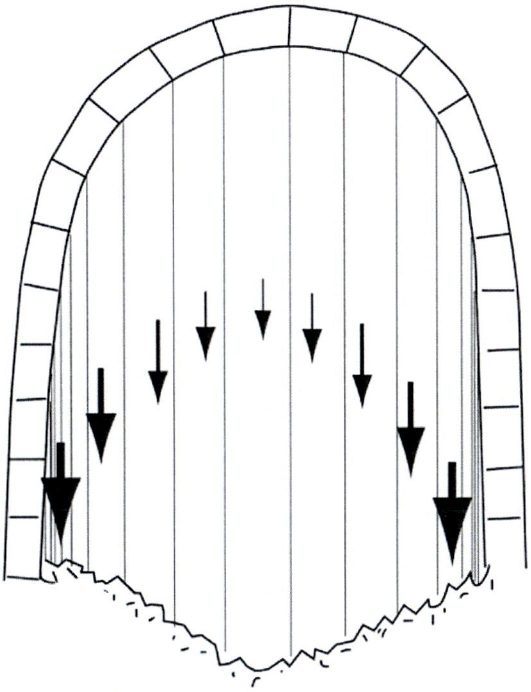

Fig. 2.94 Genesis of a false trough in Mort Ru Chamber

located; the thickness of limestone above the ceiling is about 25 m. In the entrance area, the Upper Urgonian mass was subject to an extensive collapse (Fig. 2.92). At the top of the cliff, at 1700 m altitude, a bare and karstified limestone plateau extends. It is situated above the Hauterivian marls.

The Mort Ru Cave is located on the western flank of the Cretaceous Seuil syncline. This unit is intersected by an important N50° vertical fault that extends near the chamber, as well as by a median N10° vertical fault and several N50° transverse faults (Fig. 2.93). The hall is dug at the base of the "Lower Urgonian Mass" in red sandy limestone aged Lower Barremian. This limestone is stratified in small beds about 10 cm thick; they have an N0° strike with a dip of 5° to the east. Fracturing is very dense mainly in the upstream part of the chamber. This one is "chopped" by a very tight bundle of fractures whose more frequent direction is N90°. An N48° fault cuts the entrance dome.

The floor of the room is completely covered by a scree with blocks 20 cm in diameter. Some several meters large blocks occupy the central part. At the foot of the southwest and northeast ends of the entrance dome, two scree cones are

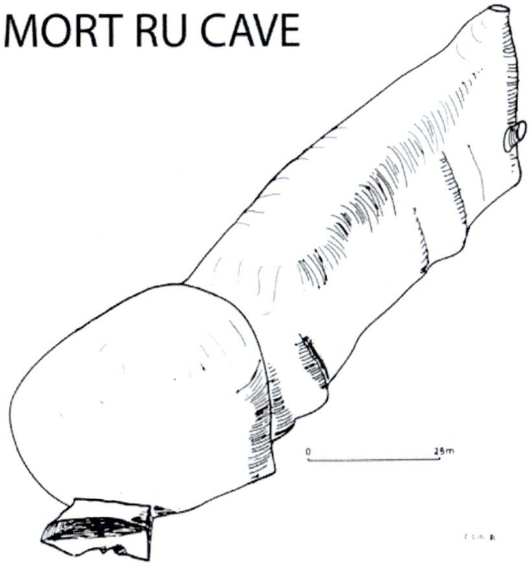

Fig. 2.95 3D sketch of Mort Ru Chamber

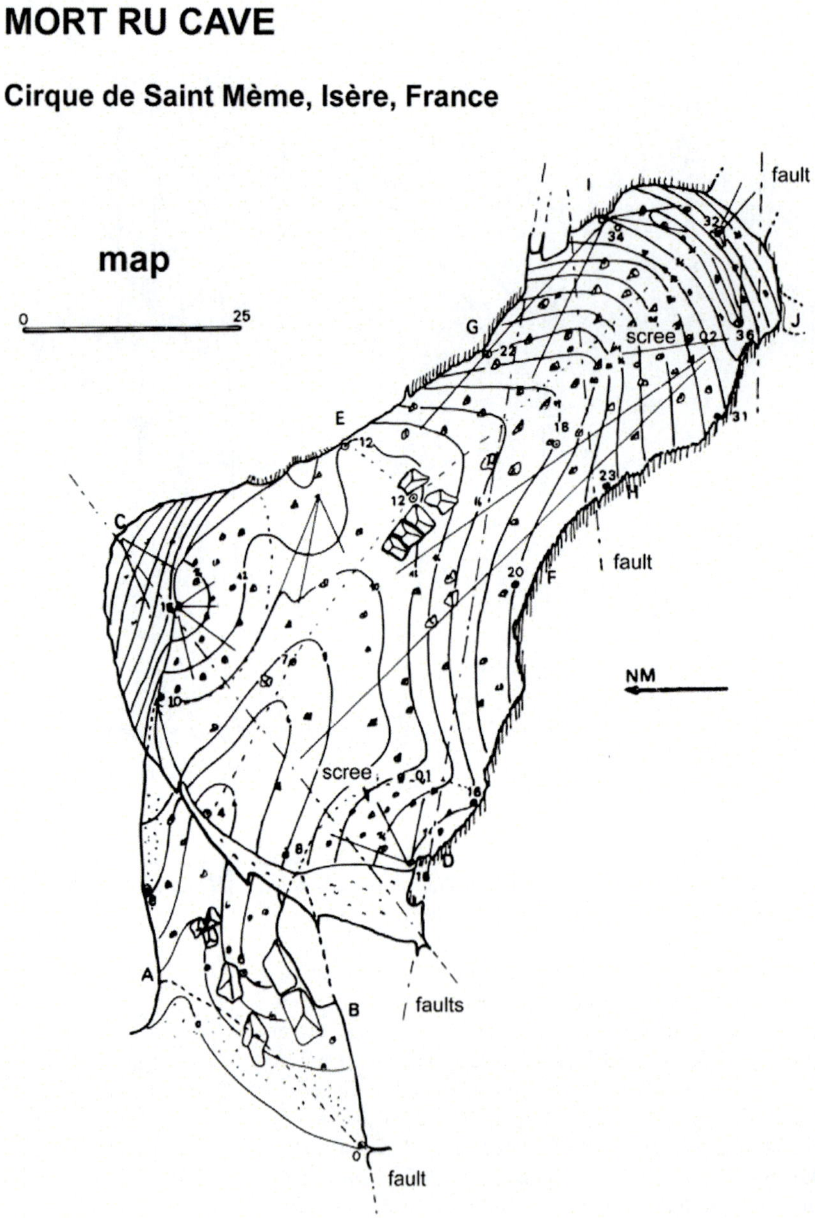

Fig. 2.96 Map of Mort Ru Cave

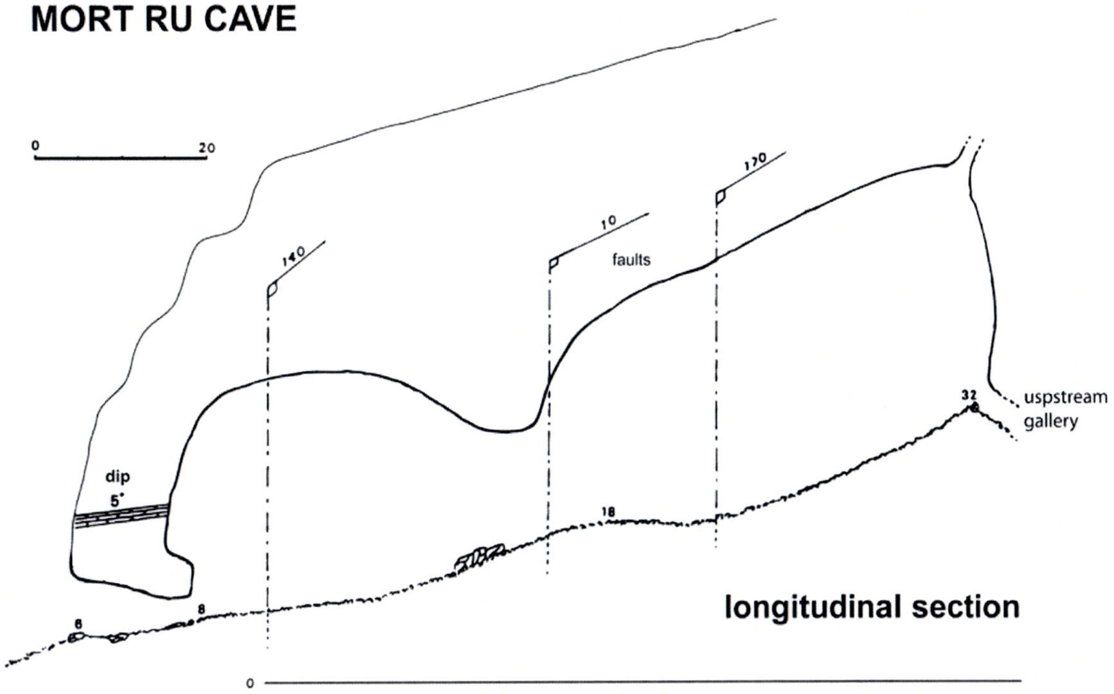

Fig. 2.97 Longitudinal section of Mort Ru Cave

present. They are formed of blocks of smaller size (10 cm in diameter). These screes are regularly fed by stones that are permanently falling from the ceiling (one stone per hour during the visit).

The presence of a trough in the axis of the chamber is not related to a water circulation or to the existence of a withdrawal. It is the meeting area of two screes supplied by the blocks falling from both walls and the ceiling. For the same amount of limestone falling from the ceiling or the walls, the receiving surface widens from the walls to the center (Fig. 2.94). The room is totally dry. It must be pointed out that there is a spring in the thalweg below the entrance of the cave. It is perhaps related to the cave.

2.13.3 Genesis and Stability

This room is a former resurgence that has enlarged near the cliff in a highly fractured area in the meeting zone of a network of wet galleries and a network of dry ones. Currently, the room seems to be in the phase of profile adjustment. The rock is disorganizing, and no mechanism ensures the evacuation of the debris. All around the chamber, the walls show a significant scaling phenomenon, especially at the southwestern wall. Rockfalls are frequent (one per hour). However, the vaulted ceiling appears quite stable. Below this vault, the cave is still evolving by a gradually crumbling of the decompressed zones.

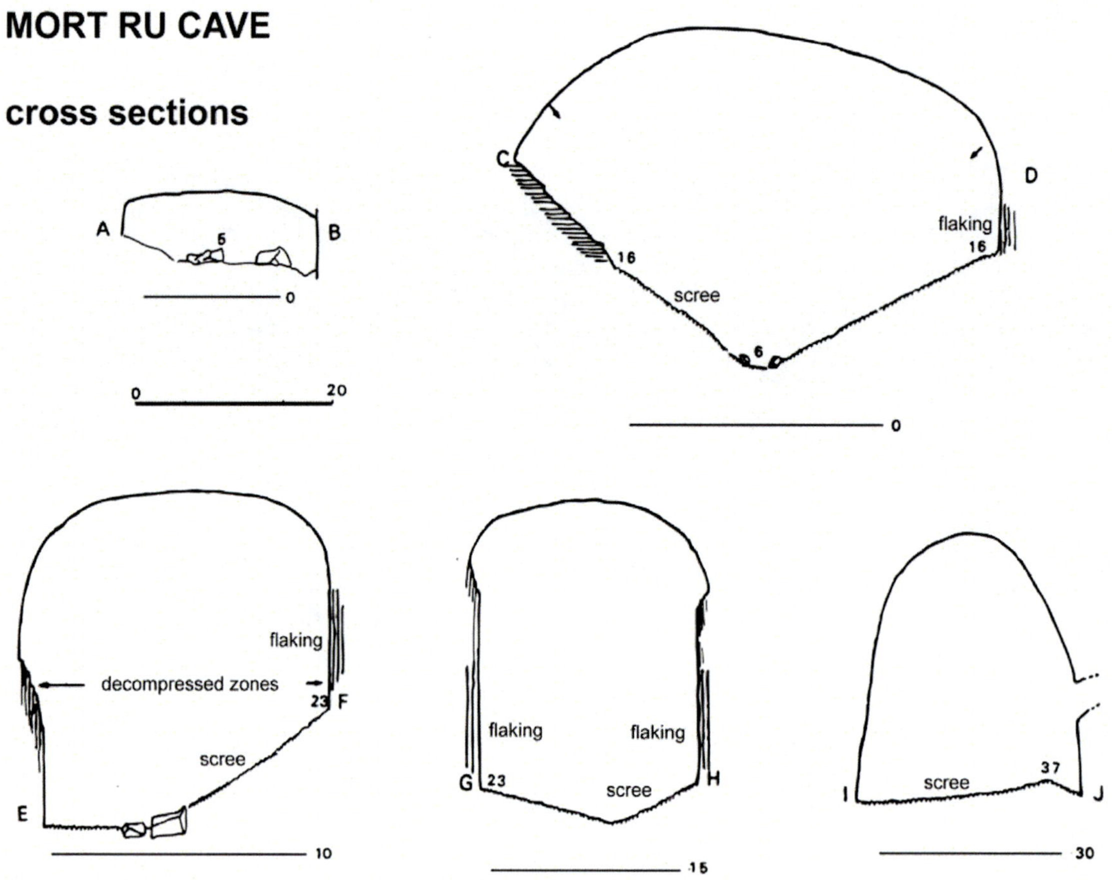

Fig. 2.98 Cross sections of Mort Ru Chamber

2.14 Aven Noir (Nant, Aveyron, France)

2.14.1 Description

See Figs. 2.99, 2.100, 2.101, 2.102, 2.103.

The Aven Noir (Black Shaft) is locally very well known. Access is free except for a recently discovered part of the cave that is protected by a door (Cavers et al. 2007). From Nant take the D991, then the D145 toward Trêves. After Cantobre drive for 2.6 kms along the Trévézel Valley, on the right bank, there is a small valley called Valat-de-Long-Bedel Valley. The chasm opens on the right bank of the Valat. Leave the car, descend to the bottom of the Trévézel Valley and reach the Valat-de-Long-Bedel. A well-marked trail leads to the shaft entrance at the foot of dolomitic limestone ledges (coordinates: 44° 4’ 36.03” N; 3° 19’ 14.71” E).

The entrance pitch leads to the ceiling of the Entrance Hall, a room 40 m in diameter. Single-rope technics are necessary to climb down the 37 m pitch and reach the summit of a scree slope at the entrance of a gallery inclined 25° to the east. This gallery is 250 m long. In the beginning, the width is 17 m and reaches 70 m at the end in an area called Balsan Hall. The bottom of this part is 107 m below the entrance which corresponds to the altitude of the Trévezel River. The ceiling of the room is flat, and the walls are vertical. This gives a rectangular shape to the sections.

2.14.2 Geology and Hydrogeology

The Trevezel Valley cuts the southern edge of the Causse Noir Plateau. The karst river has several penetrable ponors. During dry periods, the entire flow is swallowed. The slopes are dug in alternations of thick dolomitic limestone beds and softer marly levels. The river has a fairly regular V-shaped profile interspersed with five rocky ledges. The shaft opens at the foot of the third ledge between two dolomitic spurs

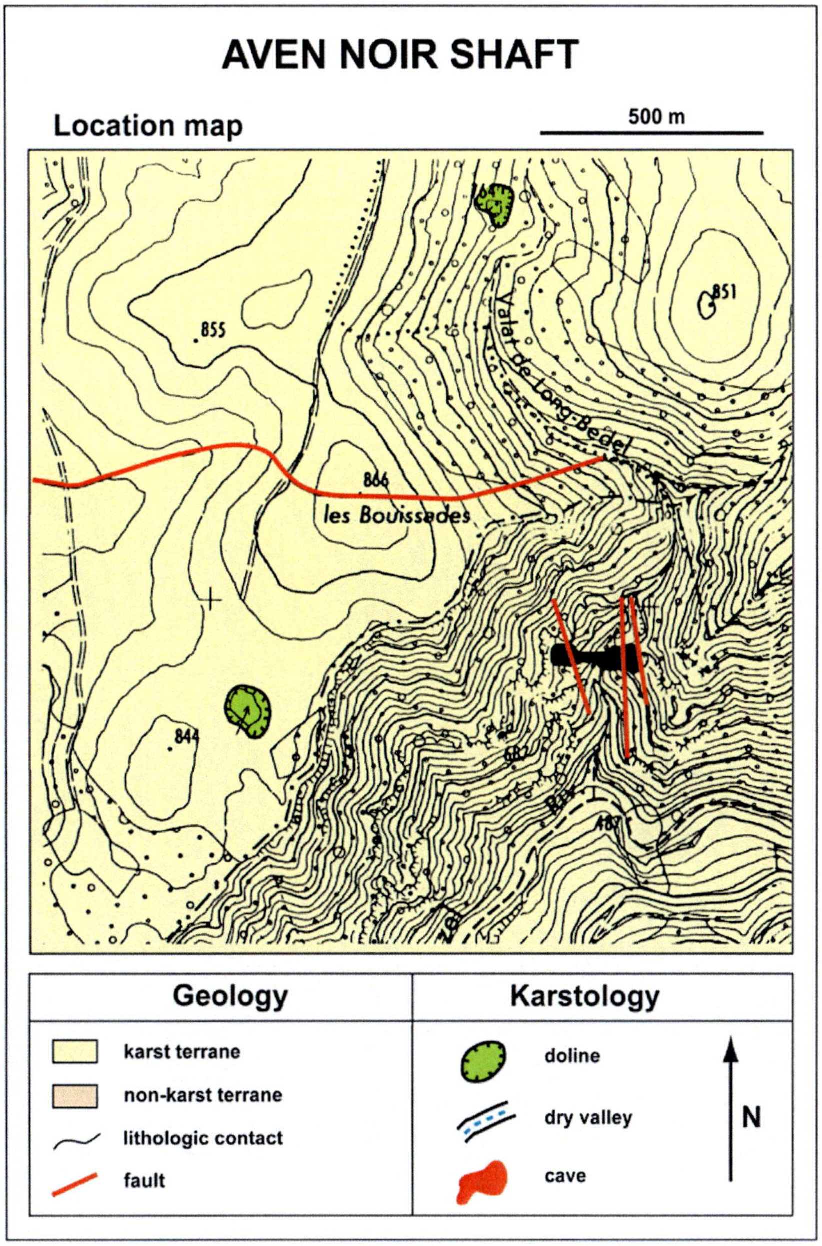

Fig. 2.99 Geomorphological map of Aven Noir Shaft

that belong to the second ledge. The cave extends to the east and passes below the Valat-de-Long-Bedel. At this level, the thickness of limestone above the ceiling is 15 m.

The cave extends in horizontal carbonate series of the Causse Noir Plateau. The aquiclude is constituted by Liassic marls. The main directions of the fracturing are W-E and SN-NE. The first direction is related to a compression phase, while the second one is related to a distension phase (Gèze 1965).

From the entrance pitch to the back of the room, the lithology is as follows:

- an Upper Bathonian dolomitic mass that forms the ceiling of the Entrance Hall;
- limestone in small beds, in which most of the entrance hall develops;

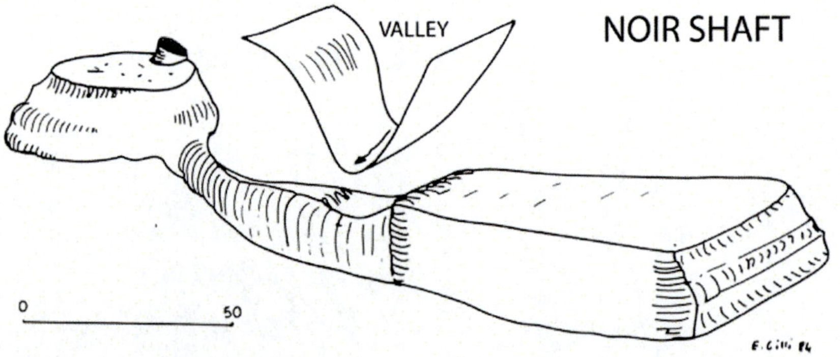

Fig. 2.100 3D sketch of Aven Noir Shaft and Valat-de-Long-Bedel Valley

Fig. 2.101 Entrance shaft of Aven Noir. Note the contact between the dolomite and the limestone in small beds

- both series surmount three large beds of dolomite which form the ceiling of the Balsan Hall;
- this chamber is entirely dug in sublithographic limestones in small beds that part in plates.

The three latter formations belong to the Lower Bathonian, whose base is constituted by a marly and lignite set that forms an impermeable level of reduced thickness between the Bathonian and the Bajocian. It constitutes a local aquiclude for the groundwater. The Balsan Hall extends from west to east according to the limestone dip which is here of 5°. The strike is N0°.

At the bottom of the room, between two N6° and N170° vertical faults (see longitudinal section E.F—Fig. 2.103), the layers have an anticline structure. Part of the eastern wall of

AVEN NOIR SHAFT

Nant, Aveyron, France

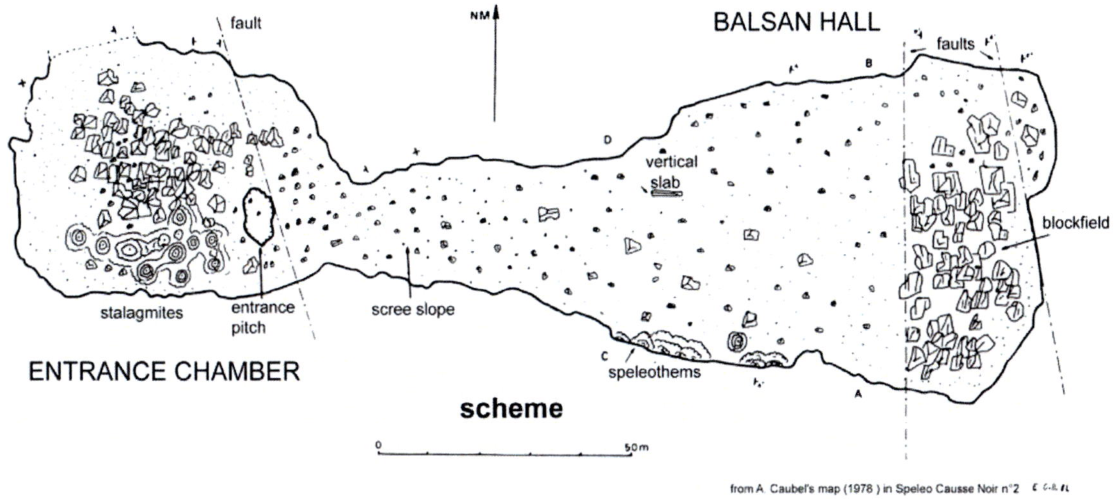

Fig. 2.102 Map of Aven Noir Shaft

the chamber corresponds to the wall of the N170° fault (Fig. 2.102). This direction is observed outside the cave on the edges of the cliffs on the left bank of the Valat-de-Long-Bedel, which presents structure planes that are oriented identically. The access gallery to the Balsan Hall is also excavated under a small anticline axis N0°. The eastern part of the Entrance Hall extends in a small syncline of the same direction, cut by an N160° fault. The entrance shaft is aligned on that fault. Near this fracture at the north wall, there is a tight network of N50° diaclases.

In its center, the floor of the Entrance Hall is occupied by a blockfield of metric-size elements. It is covered with speleothems in its southern part and more clayey in the northern and western parts. To the east, the size of the blocks decreases. They form a slope inclined at 25° which gives access to the gallery that leads to the Balsan Hall. The floor of the latter is formed of small blocks and platelets centimetric to decimetric. At the eastern end of the room, the floor is again covered by a blockfield of large parallelepipedal blocks that mainly come from the eastern part of the ceiling in a zone framed by two faults.

The Valat-de-Long-Bedel is a tributary of the Trévezel River. It is born on the Causse Noir Plateau. A sharp slope break in the longitudinal profile of this torrent, at the edge of the plateau, suggests the presence of a swallow zone that gives access to the cave network of the Aven Noir. This makes possible the connection of the Causse groundwater with the local base level after the deepening of the Trévezel riverbed.

2.14.3 Genesis and Stability

In the Aven Noir Shaft, both chambers are located on a gallery that is probably the underground riverbed of the Valat-de-Long-Bedel. This gallery has been originally dug out in a massive limestone. It enlarged and formed the Entrance Hall when reaching the small limestone beds of the Lower Bathonian to give, then the Balsan Hall, when reaching the platy limestone. The narrowing of the gallery between the two chambers occurs when crossing the massive dolomitic layers. However, the presence of marl beds at the base of the Lower Bathonian suggests a digging by scouring of the marls and then collapse of the limestone.

Apart from the existence of signs of instability in a very diaclastic zone near the entrance fault, all the walls of the cave look very stable. There may be phenomena of racking at the eastern end of the Balsan Room.

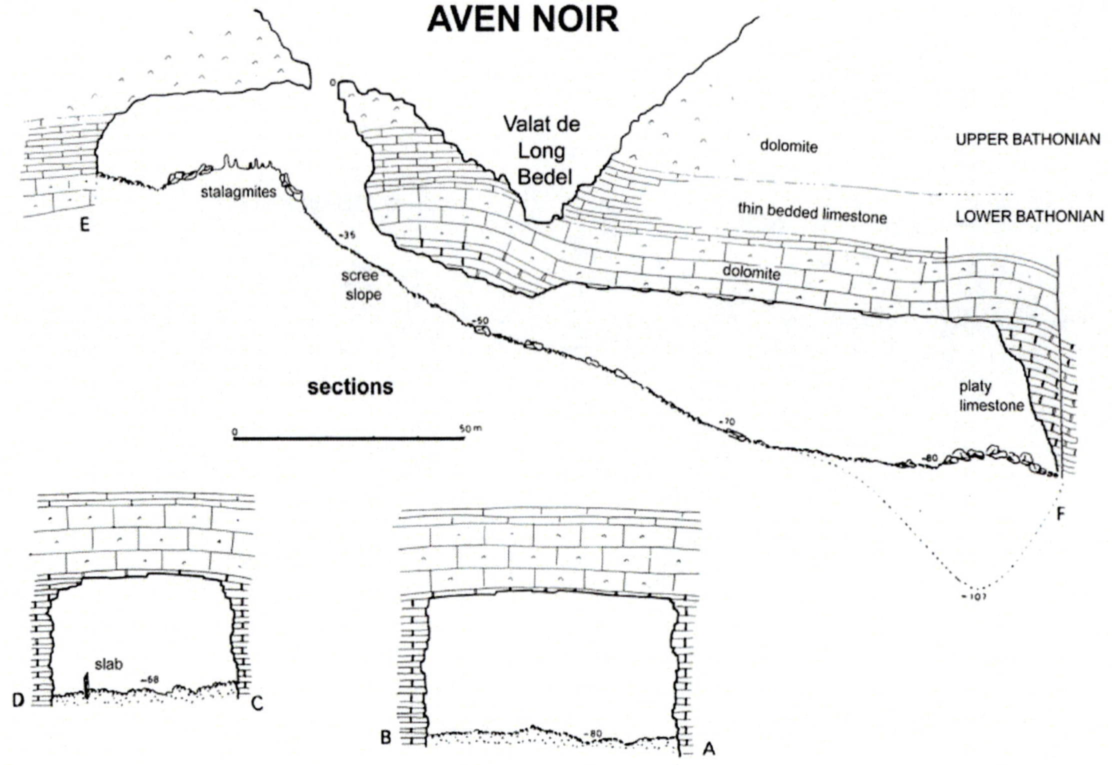

Fig. 2.103 Cross sections of Aven Noir Shaft

2.15 Aven d'Orgnac—De Joly Hall (Orgnac, Ardèche, France)

2.15.1 Description

See Figs. 2.104, 2.105, 2.106, 2.107, 2.108, 2.109.

The Aven d'Orgnac (Orgnac Shaft) was explored in August 1935 by the French caver Robert de Joly. The presence of a spectacular forest of stack-of-plate stalagmites justified the development of the cave for tourism in 1939. In 1965 and 1966, important extensions were discovered in the north by the cavers of the Speleological Center of Vallon-Pont-d'Arc. They have shown that the tourist part of Aven d'Orgnac is actually a small part of a giant gallery very rich in speleothems. Nowadays, the cave is an important tourist center that includes a museum. It is classified "Grand Site de France" (great French site) since 2004. Part of the cave is monitored for karst studies (Delannoy et al. 2007).

The present work only concerns the De Joly Hall, which is the main enlarged portion of the tourist part. Access to the show cave is by the D317 from the village of Orgnac-l'Aven, 25 km west of Pont-Saint-Esprit (coordinates: 44° 19' 11.93" N; 4° 24' 43.04" E). A tunnel, then a lift, gives a direct access to the chamber.

The De Joly Hall is oval-shaped and measures 115 m long and 80 m wide. The floor is covered with large blocks and stalagmites; it is inclined toward the southeast. The ceiling is flat in the center and descends in reversed steps toward the northern and eastern walls. It presents a recess in its center, thanks to a diaclase which crosses the whole room in its major axis (see sections AB and CD—Figs. 2.108 and 2.109).

The natural entrance is a pitch that pierces the ceiling close to the south wall of the hall. At the bottom of this well, 49 m below the surface, there is the head of a coarse scree cone made of small blocks and earth. Behind it, there is the opening of the artificial tunnels that give access to the chamber.

Fig. 2.104 Geomorphological sketch of Orgnac Shaft

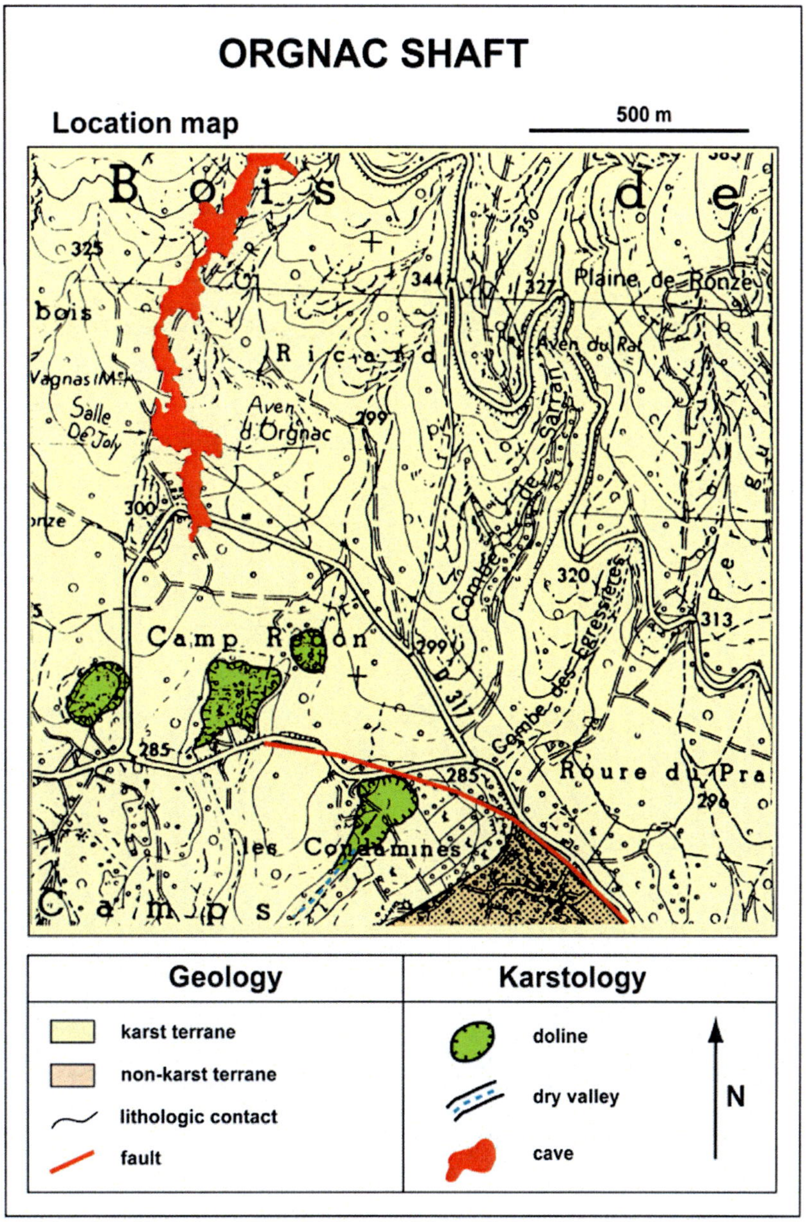

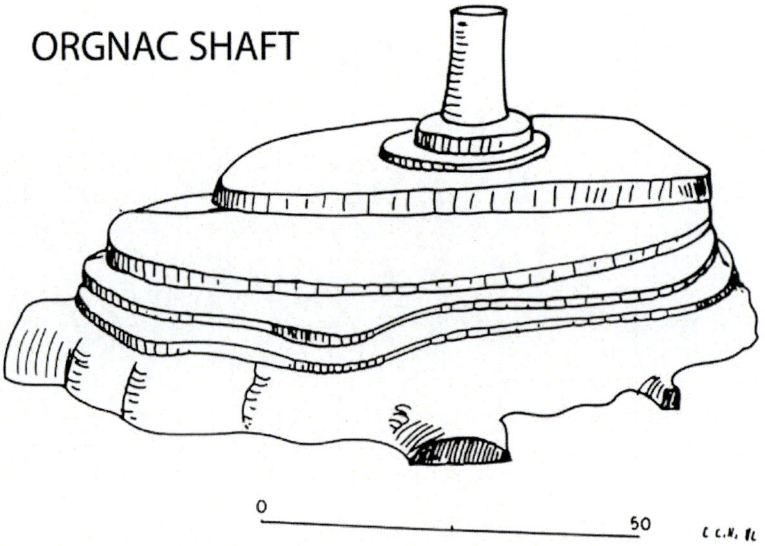

Fig. 2.105 3D sketch of Orgnac Shaft

Fig. 2.106 Stalagmite forest of Orgnac Aven. Note the entrance shaft in the background (Photo Benh LIEU SONG—CC BY-SA 3.0)

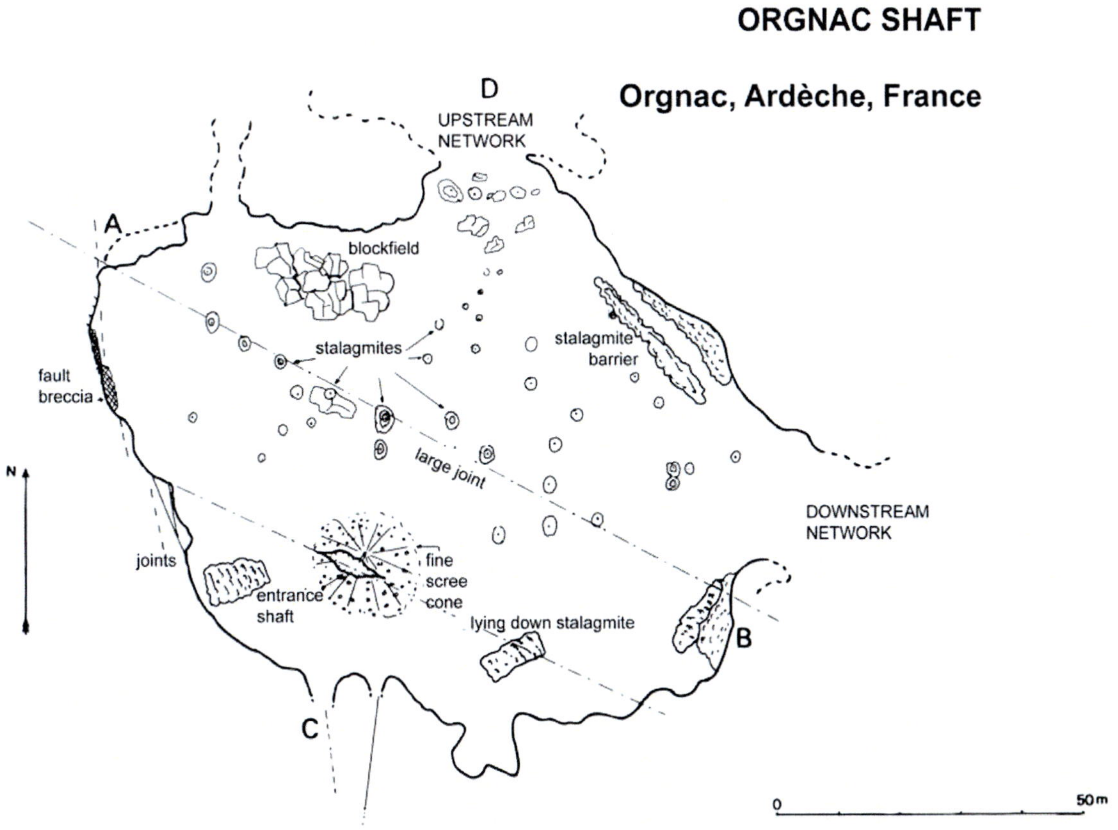

Fig. 2.107 Map of De Joly Hall in Orgnac Shaft

To the east of the room is the departure of a descending gallery, the Red Hall (southern terminus of the Aven). This gallery makes an S and goes under the east wall of the De Joly Hall. To the north, a low passage, cluttered with speleothems, gives access to the upstream network of galleries.

2.15.2 Geology and Hydrogeology

Located 305 m.a.s.l. on the karst plateau that separates the canyons of the Ardèche and the Cèze, the cave is dug in the Urgonian limestone. The cave network develops under the Bois-de-Ronze ridge which separates the Cèze River and the Ardèche River basins. This particular position illustrates the principle of relief inversion. The presence of the cave protects from superficial erosion the terrain that is situated above it. Consequently this highlights it in the landscape (Gèze 1969). Above the De Joly Hall, the limestone roof has a thickness of 15–20 m.

The Orgnac Shaft is located on the Ardèche Causse, a limestone plateau with a gentle dip to the east. The cave is entirely dug out in the thick series of Urgonian limestone dated from the Upper Barremian. The limestone beds are one to several meters thick. At room De Joly Hall, the layers have an N45° strike with a 10° dip to the east. An N170–180° fault with a 65° dip to the west affects the western end

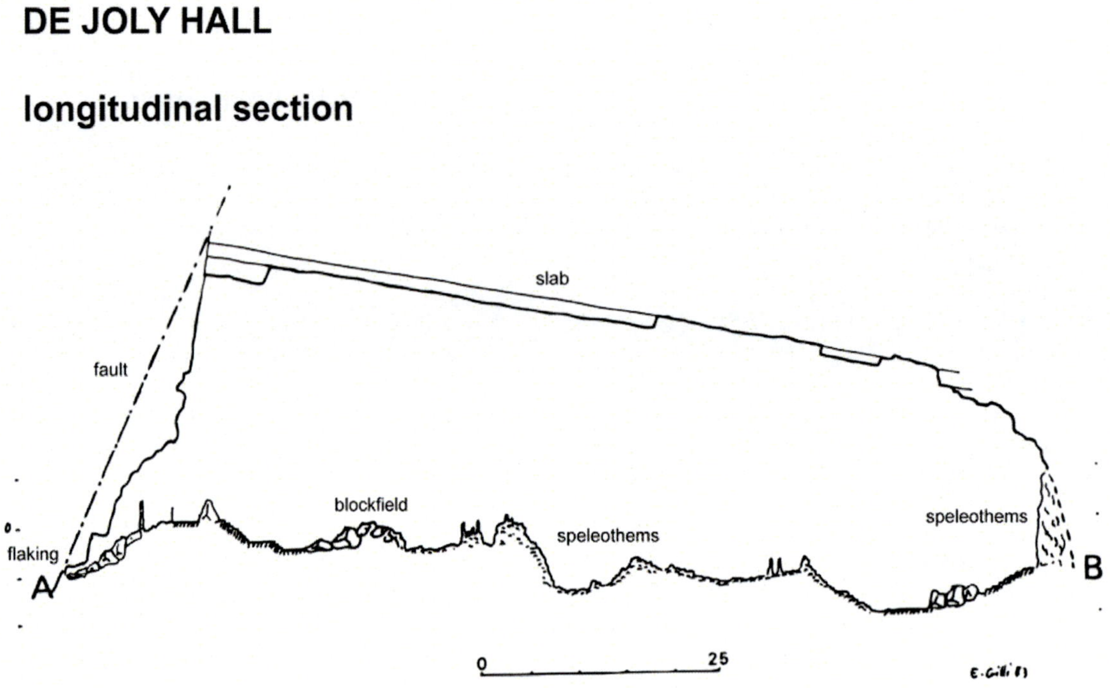

Fig. 2.108 Cross section A–B of De Joly Hall in Orgnac Shaft

of the room. The wall is disorganized and locally brecciated. Significant N120° vertical joint intersects the room. The entrance shaft is aligned on one of them.

The central diaclase caused the rupture of the limestone beds which formed a large recess on the roof (see section C–D—Fig. 2.109). Another set of diaclases is visible on the ceiling; it overlaps the previous one at 90° and has little influence on the morphology of the chamber. These joints are underlined by a significant amount of speleothems.

There are three types of filling:

The small scree cone located at the bottom of the entrance shaft. The scree is earthy and formed of small limestone blocks. The elements mainly come from outside (Figs. 2.107, 2.109).

The floor of the hall. It is made of variable size blocks, collapsed from the ceiling and constituting a blockfield.

The speleothems. They are highly developed which makes sometimes difficult the observation of the walls. They cement the blockfield that forms the floor of the room. They locally form a forest of stack-of-plate stalagmites which is the main attraction of the tourist part. To the right of the opening of the access tunnel, a large stalagmite has switched. Its fall seems linked to a differential settlement of the underlying clay.

The Aven d'Orgnac is today a totally dry cave. The network map suggests that in the past an underground river drained the cave from NE to SW.

(**Note**: In 2002, it was observed that during a very rainy period the groundwater level rose up and flooded the Salles Rouges, the lower parts of the touristic path.)

2.15.3 Genesis and Stability

The Aven d'Orgnac is actually a section of a giant gallery. Becoming dry, this gallery was then compartmentalized by important curtains of speleothems. The collapses that affected the ceiling contributed to enlarge the cave until reaching the surface which caused the opening of the entrance shaft. The important growth of speleothems at least proves the stability of the cave for tens of thousands of years. In the western end of the room, one area looks unstable and evolutionary. Indeed, the presence of an N 0° fault has favored the dislocation of the wall and rock scaling is observed.

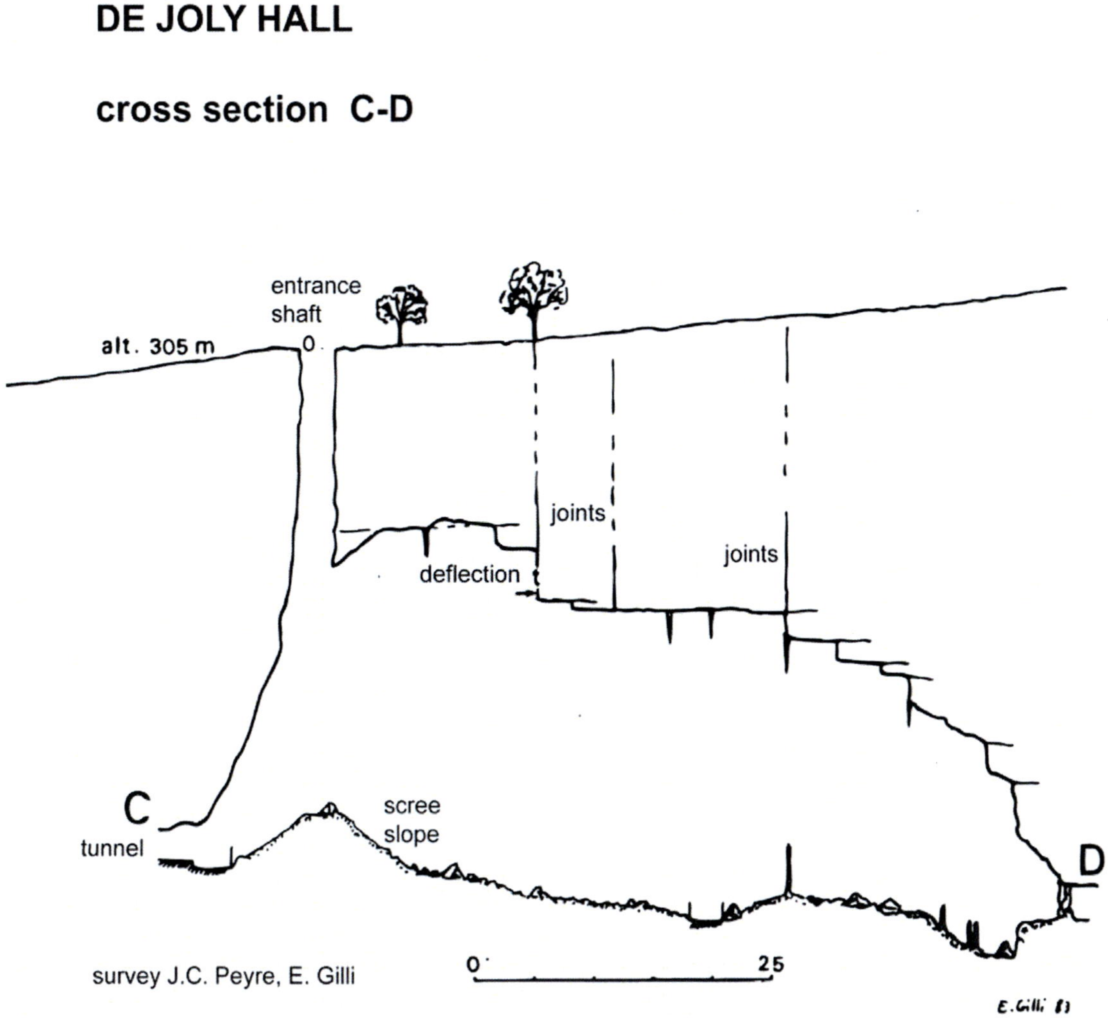

Fig. 2.109 Cross section C–D of De Joly Hall in Orgnac Shaft

2.16 Padirac Shaft (Padirac, Lot, France)

2.16.1 Description

See Figs. 2.110, 2.111, 2.112, 2.113, 2.114.

The Padirac Shaft was known due to its large dimensions. It was first explored by E.A. Martel in 1889 (Martel 1894). Exploration is still in progress. The entrance pitch and part of the cave network were developed for tourism (Dubrana 2013). The access is by Gramat driving on the N681 for 8 km toward Martel to the north, then by the D673 until Padirac village. From there, a road to the north leads to the shaft entrance which is surrounded by tourist buildings (coordinates: 44° 51' 29.64" N; 1° 45 '0.67" E). It was studied by Renault (1968).

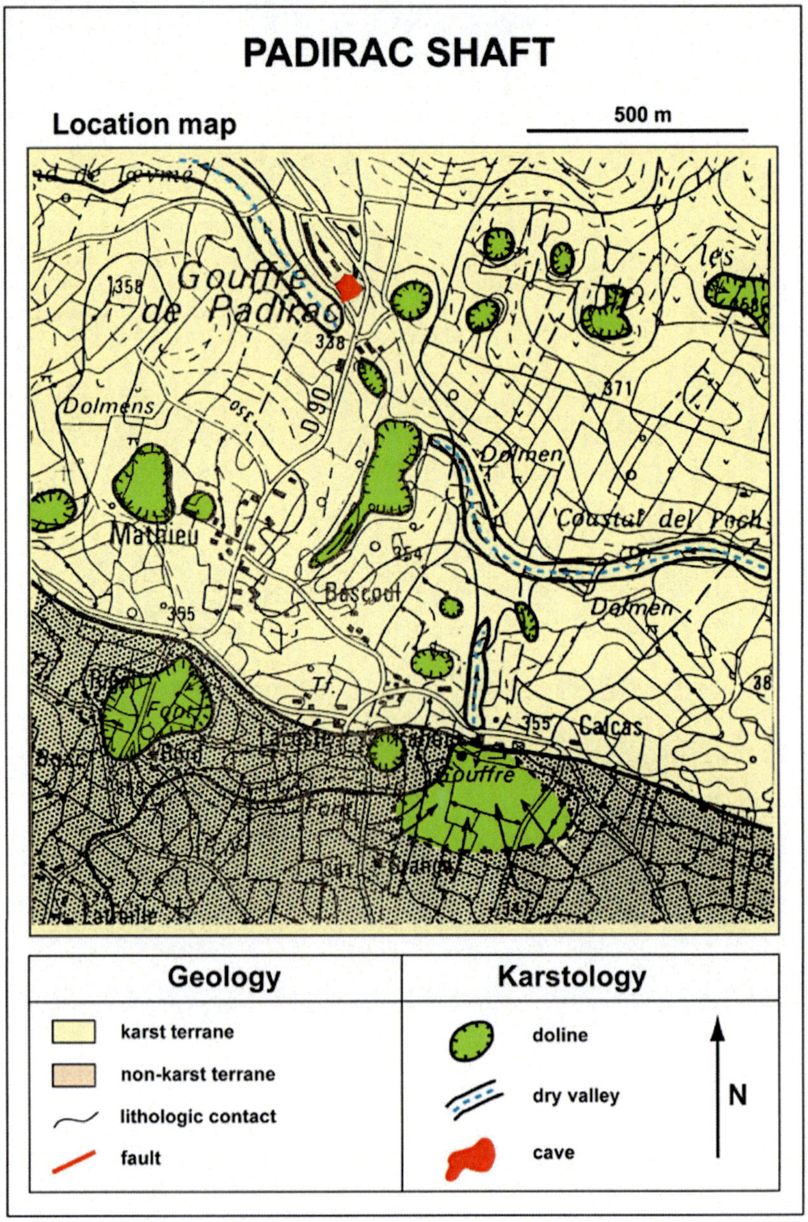

Fig. 2.110 Geomorphological map of Padirac Shaft

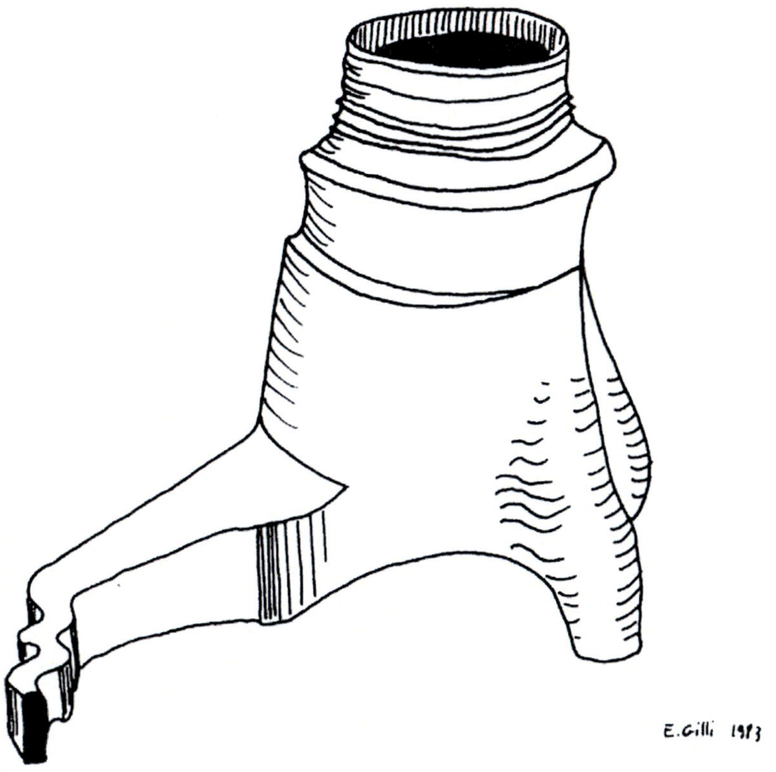

Fig. 2.111 3D sketch of Padirac Shaft

The Padirac Shaft is the perfect example of a collapse shaft. The entrance is round-shaped and 25 m in diameter. The bottom is roughly triangular with a medium width around 50 m. The depth is 45 m from the entrance to the head of the scree inclined 20° to the southeast. At its bottom, two passages give access to the underground river above which the shaft is located. Walls are vertical or overhanging. At 16 m below the surface, a large ledge was used to build a restaurant and a lift. Important deposit of tuffa and portions of small galleries are observed in that area.

2.16.2 Geology and Hydrogeology

Padirac Shaft is located 345 m.a.s.l. on the edge of a dry valley on Gramat Causse, a karst plateau with numerous dolines. This plateau is bordered to the south by a fault where limestone is in contact with Liassic marls. The marls are gullied and form small poljes whose rivers are swallowed by the limestone when reaching the fault.

The cave is dug in sub-horizontal series of Medium Jurassic limestone. The Toarcian marls form the aquiclude. They are visible south of the cave due to the presence of a significant N120° vertical fault. In that area, the limestone dip augments and plunges to the north.

From the surface to the shaft bottom, the series are:

- 30 m of lithographic limestone in large beds with a few marly interbeds; 15 m below the surface these marls beds become thicker and their sensitivity to erosion gave birth to the ledge. This formation is dated Lower Bathonian;
- 30 m of sub-oolitic limestone in hard and massive beds without marly interbeds;
- the underground river is dug in Aalenian nodulous limestone (Gèze and Cavaillé 1977).

The strike is N160°; the dip is 5° to the west. At the southeastern wall, an anticline cut by an N130° vertical fault is observed. This fault is also visible on the other wall of the shaft to the north. The north wall and the east corner are the most fractured parts of the shaft. An N100° fault inclined 48° to the north intersects the shaft in the tourist lift area. The limestone compartment between both faults is highly diaclased with two preferential directions: N130° and N85°.

Fig. 2.112 Upper part of Padirac Shaft. The corniche which is located the top of the lift and the access part extends on marly levels

The floor of the shaft is a scree made of centimetric to decimetric blocks. This scree is deposited on a blockfield of large elements. It was thus formed after the collapse of the ceiling and comes from the weathering of the shaft walls. There are no scouring signs despite the presence of the underground river. Thus, the clearing of the collapse blocks is no more in progress.

The shaft is located above an underground river, partially fed by the surface water collected by the marls depressions and swallowed by the limestone when reaching the fault. This river has several tributaries and feeds the Saint Georges spring, located to the east, close to Montvalent in the Dordogne Valley. At surface, one can see the remnant of an ancient valley on which are now aligned several sinkholes and portions of dry valleys. It is probably the previous Padirac riverbed. Most of the underground river is now explored and surveyed. In the entrance area, underground water deposited thick tuffa levels. This water circulates above the marl interbeds of the upper part of the shaft.

2.16.3 Genesis and Stability

Padirac is a collapse shaft above an underground river. It was created by the ascending enlargement of the ceiling of a portion of the gallery until it reaches the surface. A similar evolution is visible in the foremost part of the tourist path in a place called Grand Dome (Great Dome). In that place, the ceiling is close to the surface, but it did not collapse. A rupture would provoke the birth of a shaft similar to Padirac one. However, due to its shape in enlarged diaclase that narrows at its end, the roof is very stable.

Most of the digging process occurred from down to up. However, the presence of the small galleries and the tuffa deposits in the ledge area suggests that water was swallowed in some portions of the primitive surface riverbed. This caused the karstification of some limestone layers close to the surface. After the surface riverbed was abandoned, the collapse of the ceiling was favored by these brittle zones. Except for small signs of weathering similar to those of outside limestone cliffs, the shaft now looks stable.

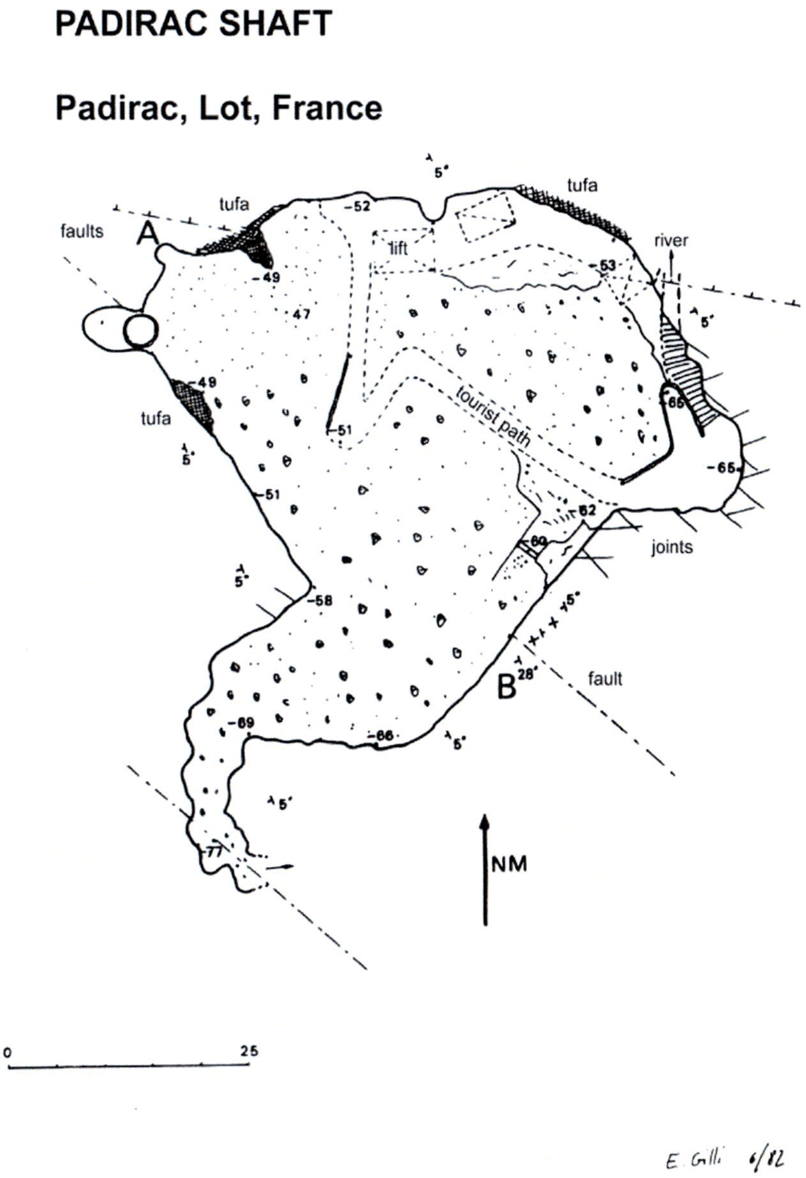

Fig. 2.113 Map of Padirac Shaft

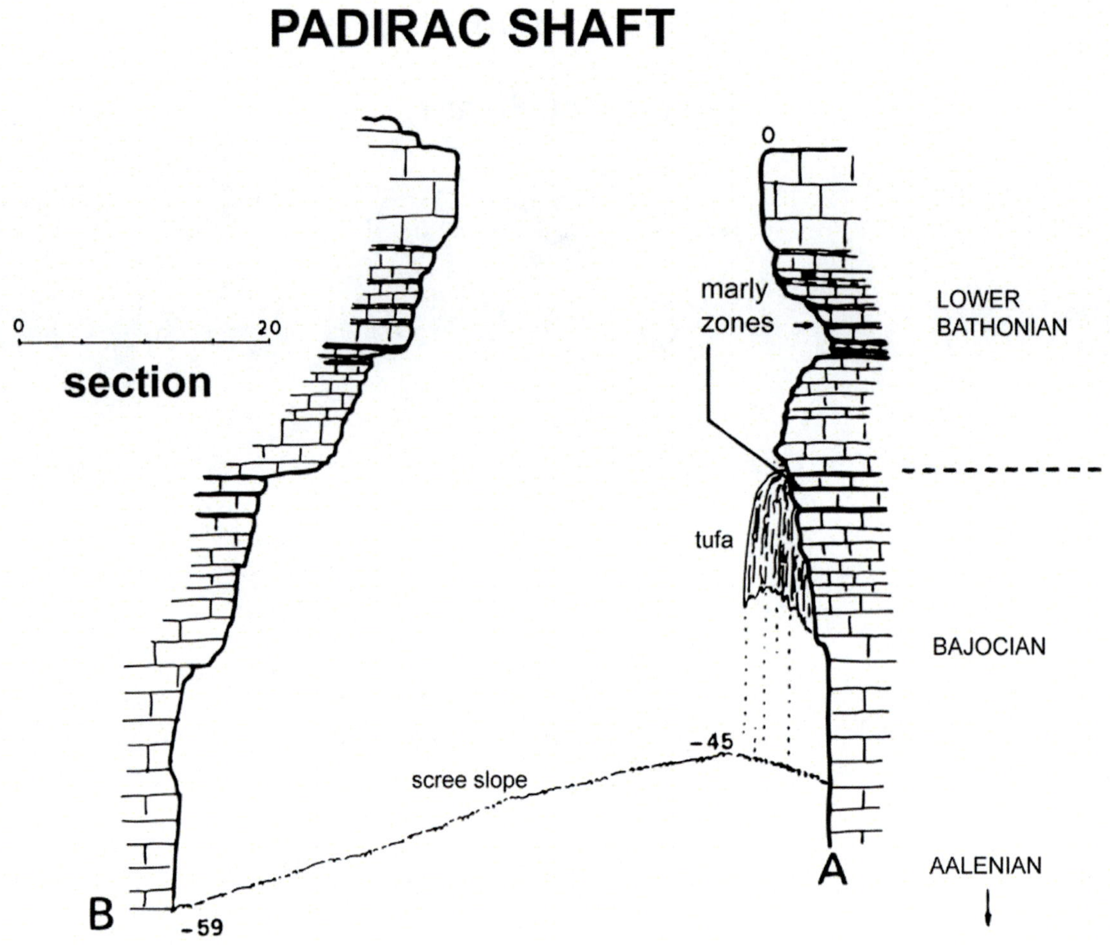

Fig. 2.114 Cross section of Padirac Shaft

2.17 La Verna Chamber (Pierre Saint-Martin Cave Network) (Saint Engrace, Pyrénées Atlantiques, France)

2.17.1 Description

See Figs. 2.115, 2.116, 2.117, 2.118, 2.119, 2.120, 2.121, 2.122.

The first entrance to Pierre Saint-Martin Shaft was found in 1950 by Georges Lépineux. The explorations started in 1952 and were bereaved by the death of a caver, Marcel Loubens. In 1953, the Verna Hall was discovered. Until 1982, it was the biggest chamber in the world. In 1953, due to its depth of 734 m, the Pierre Saint-Martin Shaft was also the deepest cave in the world. Since 1960, the chamber is easy to access using an 800-m-long artificial gallery dug for hydroelectrical purpose in the 1950s. The place is a tourist cave since 2010 and welcomes a hydropower plant (Le Bec 2008). A digital survey of the chamber was made in 2004 (Varrel 2005).

The access is from Tardets Sorholus. Drive on the D26 and then the D113 to Sainte-Engrace village. From there, a well visible forest trail leads to the entrance of the tunnel at an altitude of 1050 m (entrance tunnel coordinates: 42° 58' 45.42" N; 0° 47' 46.32" W).

The tunnel cuts the chamber at the medium height of the southern wall. The chamber is round-shaped and is 270 m long and 230 m wide (Brook 1974). The ceiling is shaped as a regular dome. It caps a scree made of schist fragments and is partially covered with huge limestone blocks. Its shape is that of a half funnel. The western part of this funnel is a mostly vertical limestone wall.

A river whose discharge may overpass 3 m^3/s crosses the chamber from one side to the other. This river was the path

Fig. 2.115 View of the bottom of La Verna Chamber

followed by the first explorers who arrived from the Lepineux Shaft located upstream. It arrives in the chamber in its medium height and runs between the large blocks before sinking in the bottom of the chamber in a pebble zone. On the opposite side of the river gallery, a few tenths of meters above the pebble zone, a large dry gallery extends: the Aranzadi Gallery. It is the previous bed of the underground river before the digging of the chamber.

2.17.2 Geology and Hydrogeology

The area is a typical high mountain karst. It is very fractured. Above 1500 m altitude, the vegetation is scarce; the limestone is naked and is intensely carved by karrens (Arre d'Anie, Arre Planère, Arre Soumcouye, Col de la Pierre Saint-Martin). Endokarst forms are significant. Several tenths of kilometers of galleries have been explored in that area. The Pierre Saint-Martin Cave network is one of the main drains of the regional karst groundwater. The Verna Hall is located below an ancient glacial valley: the Arphidia ravine. The bottom of the ravine is at 1270 m altitude; the thickness of limestone above the chamber is around 100 m (Fig. 2.120).

The Pierre Saint-Martin Cave network extends in two distinct geological formations. Indeed, most of the galleries are dug into the thick series of Campanian to Turonian limestone locally called "calcaire des canyons" (Canyons' limestone). This limestone is discordant on a Paleozoic substratum which was tectonized, folded then eroded before the deposit of the limestone. It is mainly constituted by an impermeable schist formation that is an aquiclude for the karst groundwater. Meanwhile, inside this formation some limestone series are also present and karstified which makes possible a deep circulation of the karst water inside the substratum.

The Verna chamber is located at the contact between the Canyons' limestone and its Paleozoic substratum which contains schist on the eastern part of the chamber and limestone on the western part. The substratum is intensely folded, and the layers are mostly vertical (Fig. 2.120). The ceiling is totally dug in the Canyons' limestone, while the walls and the floor are in the substratum.

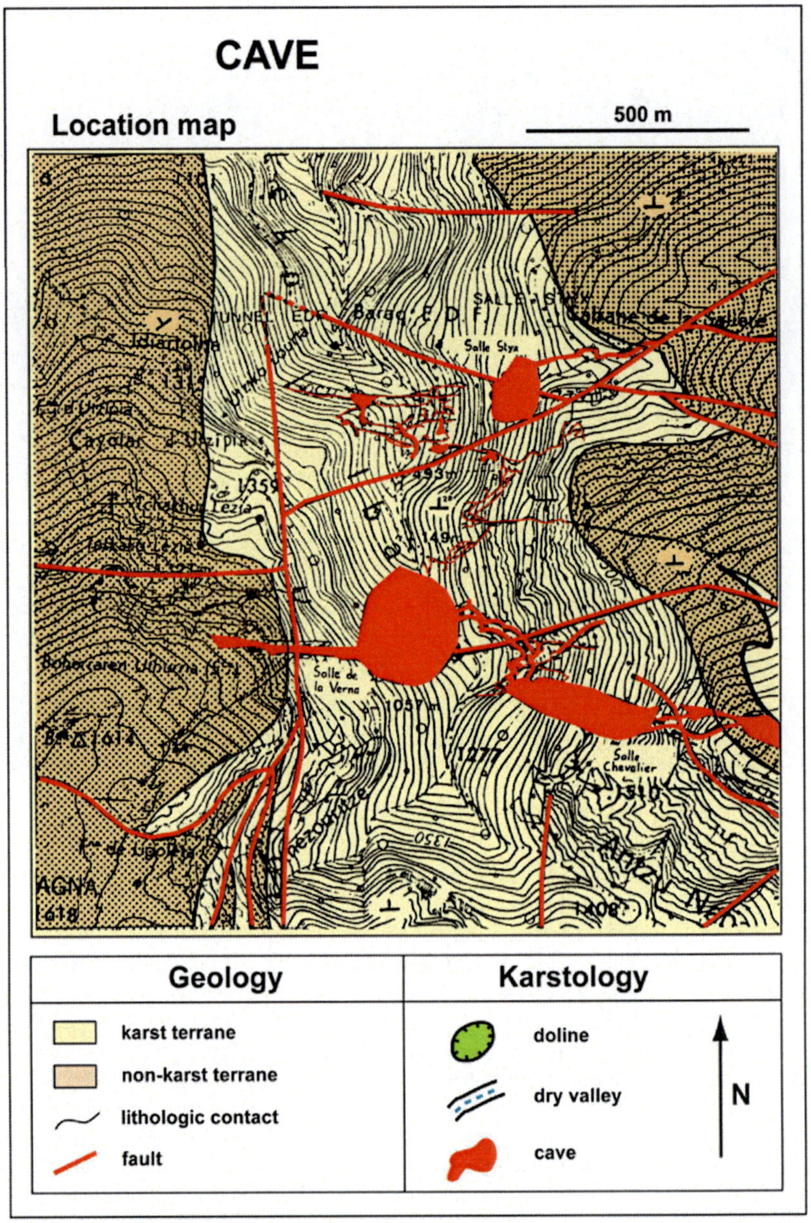

Fig. 2.116 Geomorphological map of la Verna Chamber area

Fig. 2.117 River and Paleozoic substratum of La Verna Chamber

(a) Canyons' limestone sedimentary cover. From top to base:

- 30 m calcschist and dark sandy limestone;
- 400 m massive, gravelly or micro-clastic gray limestone, dated Santonian;
- 30–50 m black massive limestone or gray crystalline dolomite with Coniacian corals and rudists;
- 15–20 m cryptocrystalline black limestone dated Turonian, this level being very reduced in that area;
- 10–15 m dark sandy limestone, sandstone and conglomerate at the basis, dated Cenomanian.

The ceiling of the chamber is entirely dug at the basis of the series up to the Santonian.

(b) Paleozoic substratum. From top to base:

- 500 m schist and Psammite Sandstone dated Namuro-Westphalian. The eastern floor of the chamber is dug in that formation;
- 100 m black limestone with white calcite veins either in thick bed or thin and corrugated layers. This formation is dated Namurian and crops out in the western part of the chamber;
- 100 m slate schist dated Visean;
- 15 m schist and lydiennes (radiolarite) dated Lower Visean and Tournaisian;
- 60 m cherry limestone with Goniatites dated Famennian;
- 50 m sandy schist dated Frasnian;
- Givetian limestone that is observed in the Arphidia Cave;
- a thick series of schist dated Lower Devonian to Silurian.

One can note the alternation of impermeable schist and karstifiable limestone. The ceiling is totally dug in the Canyon's limestone series whose dip is low. The Cenomanian and Turonian form four beds that are clearly visible at the foot of the vault and underline the contact with the Paleozoic substratum.

The upstream river circulates on the impermeable Namuro-Westphalian Schist and Sandstone. Then the river plunges into the substratum excavated by mechanical

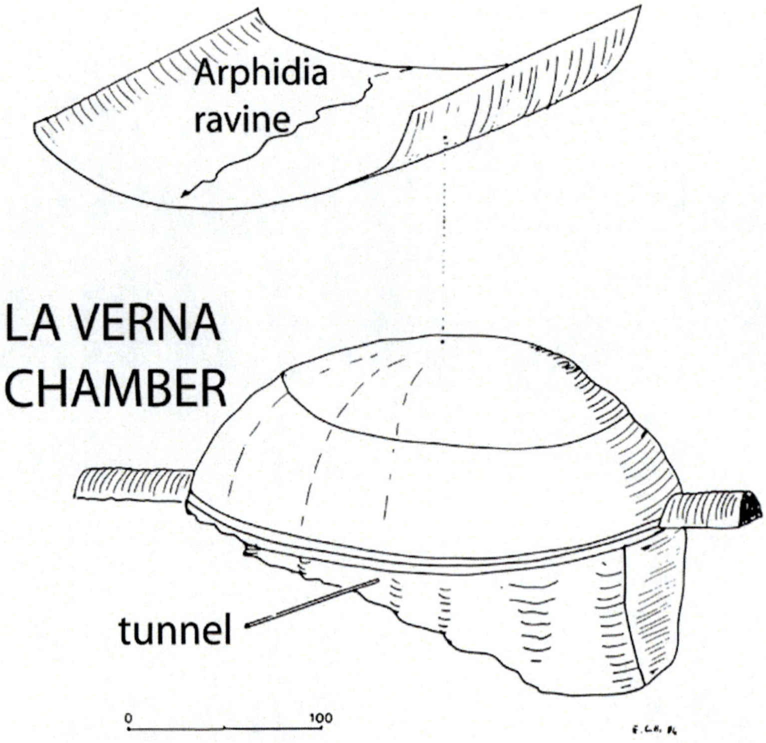

Fig. 2.118 3D sketch of La Verna Chamber

erosion. On the opposite side of the chamber, the Aranzadi Gallery is dug at the contact between the Canyons' limestone and the substratum.

In the chamber, the substratum layers strike is NW-SE with a 45–80° dip to the east. Three-quarters of the floor are dug in the schist; the western quarter extends in the Namurian limestone that also forms the western wall of the chamber. The floor of the Aranzadi Gallery is probably made of Namurian limestone at its beginning then of schist further. Several faults are visible in the chamber. The most significant one crosses the ceiling from Aranzadi Gallery to the artificial tunnel entrance. It is vertical, the throw is 2 m. Into the tunnel, the contact between the sedimentary cover and the Paleozoic substratum is very tectonized which indicates that this lithologic contact was also a slide plane.

There are three types of cave filling:

The schist talus. They are scree talus, whose elements come from schist and Psammites Sandstone. The eroded schist produces a heterogeneous earthy formation that contains small blocks. The mother rock is well visible in the higher parts at the foot of the vault. Decompression and creeping disaggregate the rock. These schist talus concern most of the chamber.

The big blockfields. The big blocks are limestone ones. They are deposited over the schist talus. Their size is metric to decametric. The biggest ones are located along the river and at the foot of the western wall. They come from the ceiling. Indeed, one can find Cenomanian conglomerate, Coniacian black limestone with Rudists and Santonian light gray limestone. They also come from the substratum like the ones that are present at the foot of the western wall. They are formed by black limestone with white calcite veins.

The sand and pebbles beach. It extends in the lowest part of the chamber in its northwestern part. The pebbles come from the erosion and the transport by the river of the previously described elements.

The river is swallowed in the substratum before reaching the pebble beach. It reaches it only during the high-water period. The Verna Hall is crossed by a river: Saint Vincent River. The flow rate varies between a few hundred and 20,000 L per second. Upstream of the chamber, the underground course of the river was explored for several kilometers, until the Arres-d'Anie, 5 km southeast of the room. The river leads into the room, cascading; the riverbed is very rough. Finally, the water is swallowed between the blocks. The Aranzadi Gallery was the ancient riverbed. It is now a

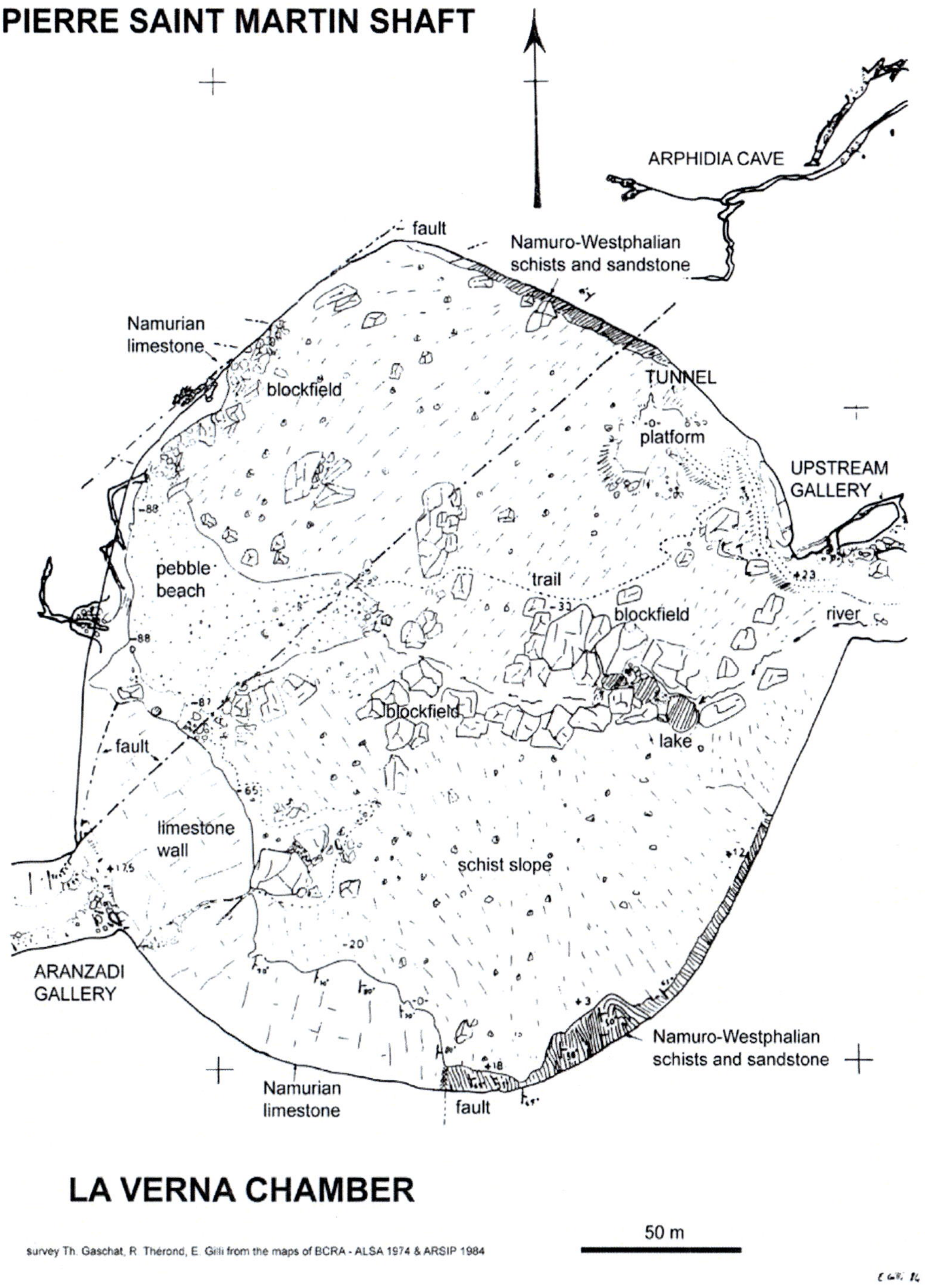

Fig. 2.119 Map of La Verna Chamber

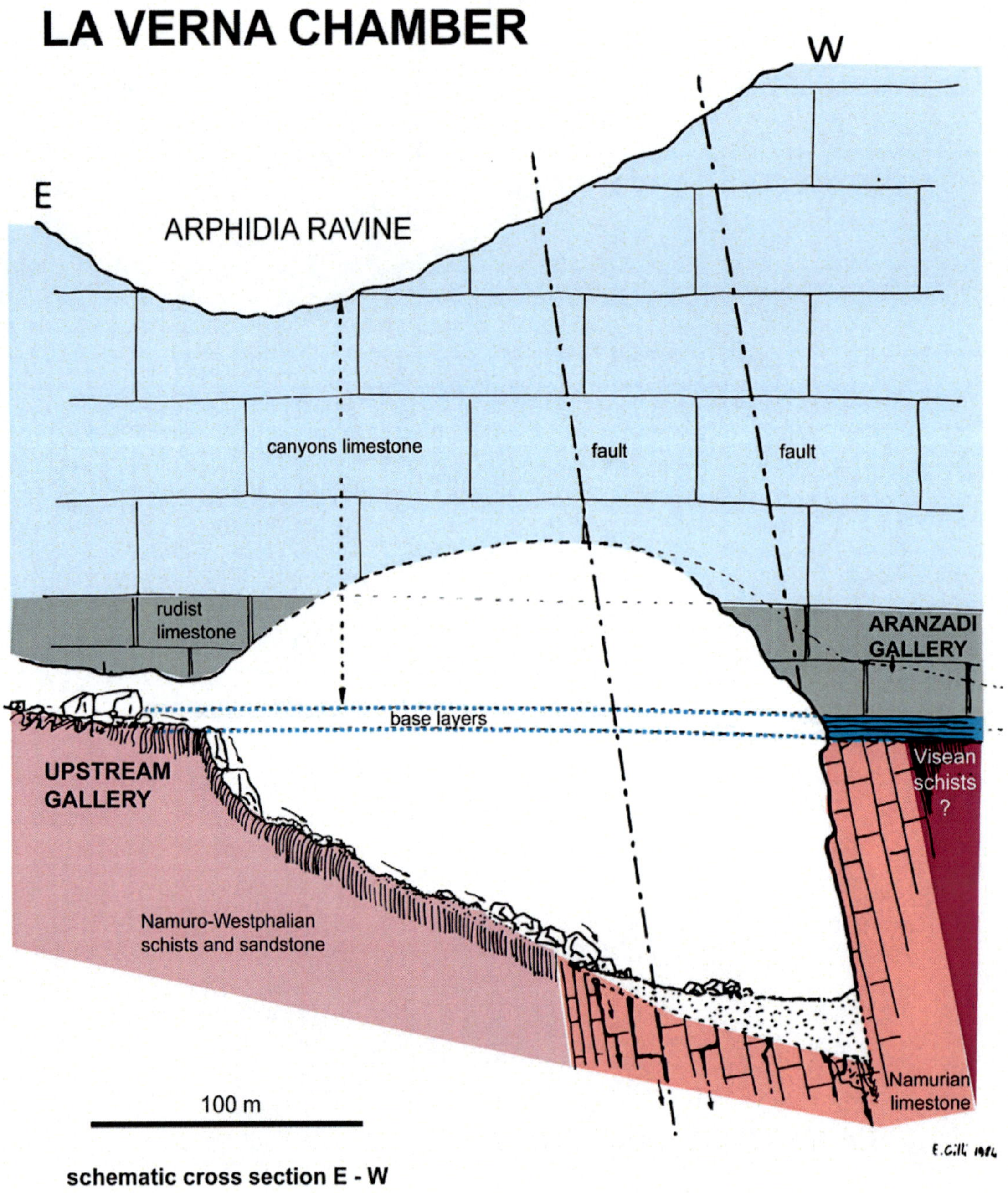

Fig. 2.120 Geological section of La Verna Chamber

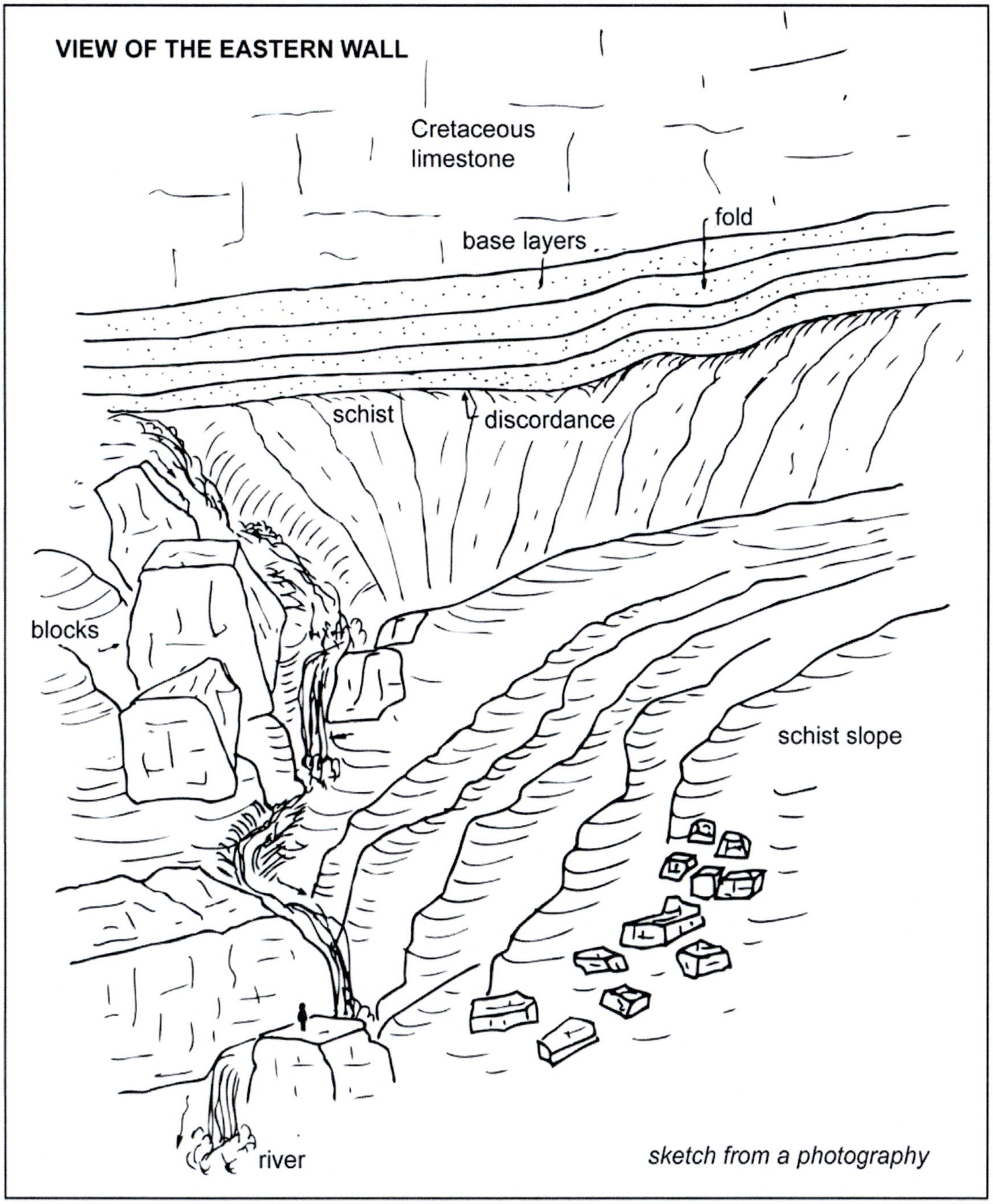

Fig. 2.121 Sketch of the lithological contact between the Paleozoic substratum and the Cretaceous limestone

Fig. 2.122 Genesis of La Verna Chamber

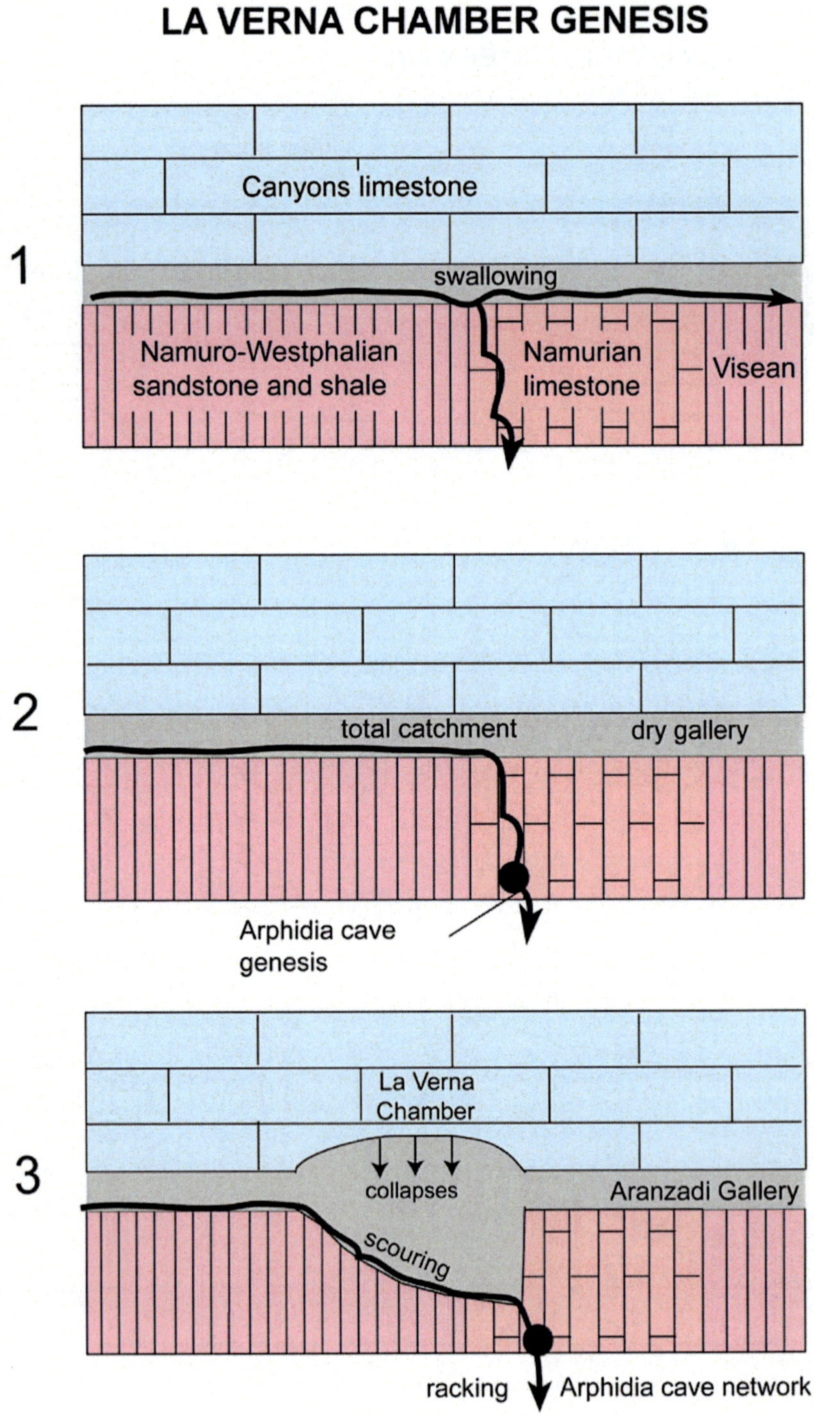

dry gallery due to the sinking of the water into the substratum. A downstream part of the underground river runs in the Arphidia Cave which was discovered during the digging of the artificial tunnel that leads to the chamber. The resurgence of the waters is now the Bentia vauclusian spring, on the right bank of the Saint Engrace dam, located in the Uhaitza Valley, about 5 km northwest of the chamber.

2.17.3 Genesis and Stability

La Verna Chamber is located on the path of the St. Vincent underground river flowing from the upstream gallery of the Pierre Saint-Martin network to the Aranzadi Gallery. This was abandoned in favor of the lower galleries belonging to the Arphidia Cave that extends in the Paleozoic basement. The passage to these galleries was made possible by the presence of the Namurian limestone visible in the western part of the room. This limestone unit formed part of the floor of the Aranzadi Gallery between the upstream Namuro-Westphalian Schist and the downstream Visean Schist. The river was swallowed when arriving in contact with the limestone. It has gradually scoured the Namuro-Westphalian Schist, and the fine fraction of the cuttings was evacuated downstream by the water running in the Arphidia galleries. The racking of the schist has created cantilevers causing the successive collapse of the walls and the vault until the present state. The Verna room is therefore a racking room, which is still active. The particular position of the room, below the Arphidia ravine (Figs. 2.118, 2.120), is similar to that of the Styx room at the bottom of the Lonné Peyré Shaft (Fig. 2.116). There is certainly a relation between this surface form and both underground rooms without it being possible to define it exactly.

At the foot of the vault, in the Cenomanian limestone and sandstone, the rock is very chipped and easily crumbles. The underlying schist is disorganized and shows some forms of creeping. The observation of the speleothems growing on the big blocks on the schist slope indicates that these elements are gently moving toward the bottom of the chamber. Some of them are unstable and they easily tilt. The room is still in digging phase. Despite its huge dimension, the ceiling dome looks stable. During our visit, a very loud explosion during a clearing session by speleologists has shaken the cavity without causing any fall of rocks. Similarly, cavers who were present in the room during the Arette earthquake (Ms 5.5, 1967) observed no rockfalls (Minvielle 1967).

2.18 Barrenc Du Pla de Périllos (Pla de Périllos Shaft) (Périllos, Pyrénées-Orientales, France)

2.18.1 Description

See Figs. 2.123, 2.124, 2.125, 2.126, 2.127, 2.128.

From Opoul-Perillos, 10 km north of Rivesaltes, follow the small road to Périllos; then 2.5 km after the Salveterra castle, turn right onto a track located 300 m before the Cortal Lalane's farm. The shaft opens on the left, 80 m after the crossroad (latitude 42.894747°, longitude 2.867471°).

Fig. 2.123 The author climbing down the Pla de Périllos Shaft in 1982

Fig. 2.124 Flat clayey floor of Pla de Périllos Chamber

A 37-m-deep vertical pitch leads to the northern edge of a circular room 75 m in diameter. The floor is clayey and horizontal with many stalagmites. The ceiling is rounded. The northwestern wall is inclined at 35° and is formed by a fault. This is the same for the southwest wall which is inclined at 60°. The southeast wall is vertical and is also a fault wall. To the north of the room, east of the entrance shaft, there is the entrance of a large gallery whose floor is cut by two parallel thalwegs. It ends after about 40 m.

2.18.2 Geology and Hydrogeology

The chasm is located 315 m.a.s.l., on a plateau south of the Montoullié de Perillou summit (707 m). This flat is arid and very karstified. There are several elongated sinkholes as well as portions of dry valleys. The only cultivable lands are the sinkholes bottoms. Elsewhere the limestone outcrops and the karrens are well developed. The vegetation is shrubby. The hydrographic network is very small and consists of a temporary torrent, the Roboul, a tributary of the Agly that has many swallow holes. The entrance of the shaft is located at the upstream end of a shallow trough, a tributary of a dry valley north of the cave. The thickness of the limestone above the chamber is about 20 m.

The Barrenc du Pla de Périllos is located east of the Paleozoic Mouthoumet massif. It is dug in the Barremo-Aptian limestone of the Corbières thrust sheet in the center of the NE-SW Périllos syncline. A set of vertical N150° faults pass through the village of Perillos 1 km from the chasm. An N40° fault is visible 500 m south of the cave (Fig. 2.127).

The rock is a very massive yellow sublithographic limestone; the stratification is hardly visible except at the surface at the lapies tables. The layers have an N30° strike with a dip of 15° to the east. The entrance pothole is located in an area cut by a very tight network of vertical N73° diaclases that are visible on the walls of the pitch.

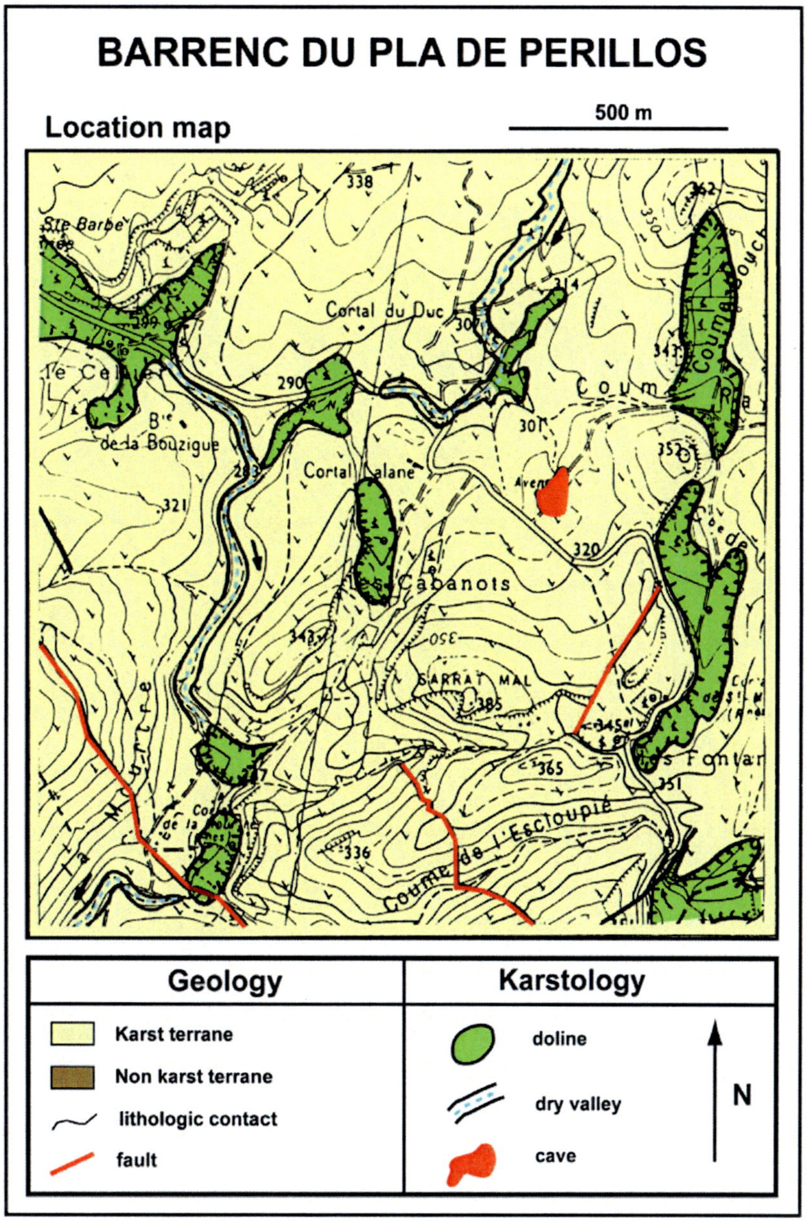

Fig. 2.125 Geomorphological map of Pla de Périllos Shaft zone

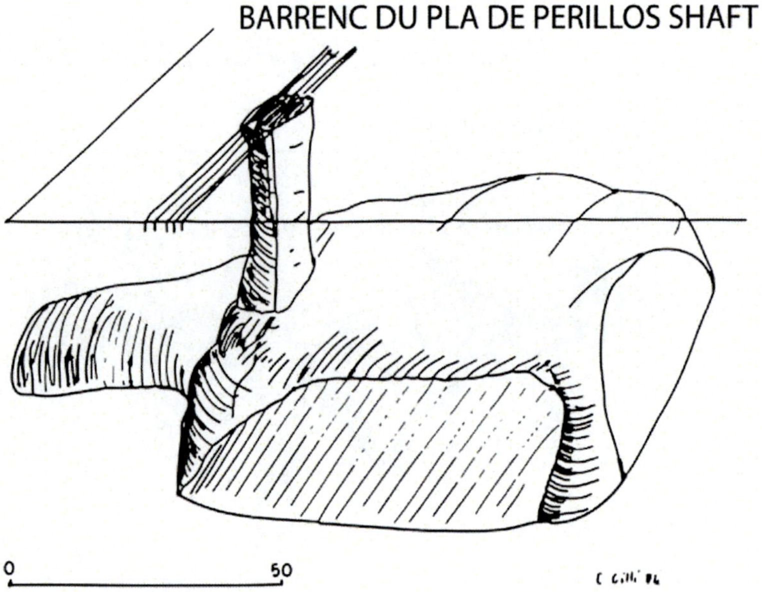

Fig. 2.126 3D sketch of Barrenc du Pla de Périllos Shaft

Inside the chamber, there are several faults. The northwest, southwest and southeast walls correspond to fault walls. The main crack directions are N30-40° and N135°. These directions are close to those observed on the surface.

The floor of the room is clayey, dotted with stalagmites, some of large size. Around the room, seepage water, forming a temporary stream during heavy rains, has dug a thalweg. The water is swallowed by small funnels south of the room or reaches the north gallery. The floor of this gallery is gullied which makes it possible to observe the different layers of the soil, the base is stony (Fig. 2.135, section AB). The cave collects seepage water that is likely to join deeper groundwater flows that feed the Font Estramar Spring located 4.5 km southeast of Périllos on the western edge of the Leucate Pond (Salvayre 1974).

2.18.3 Genesis and Stability

It is probably a chamber on a wet gallery where a vestige is visible in the north corner of the room. The water circulation is slowed down. The settling of the water gradually created a clay deposit with a horizontal surface which covered the floor of the cavity masking the blocks collapsed from the ceiling and the pebbles of the riverbed. These elements are visible on the flanks of the thalweg located in the northern part of the room. The clay deposit could also be caused by upwelling of the water table during floods.

The chamber is perfectly stable. Only the presence of racking forms in the soil of the north gallery and at the edge of the room makes it possible to envisage the existence of groundwater circulation under the cave.

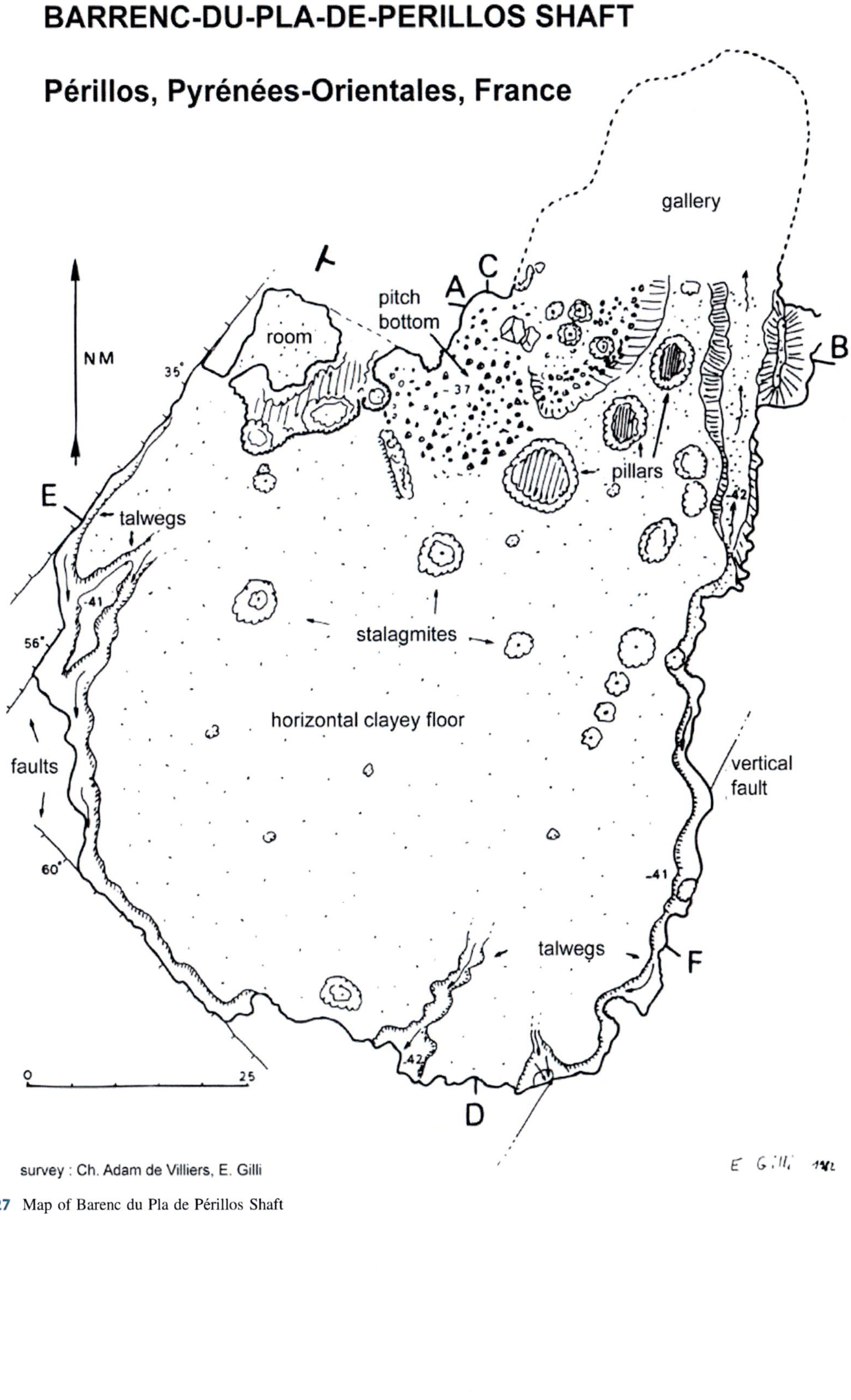

Fig. 2.127 Map of Barenc du Pla de Périllos Shaft

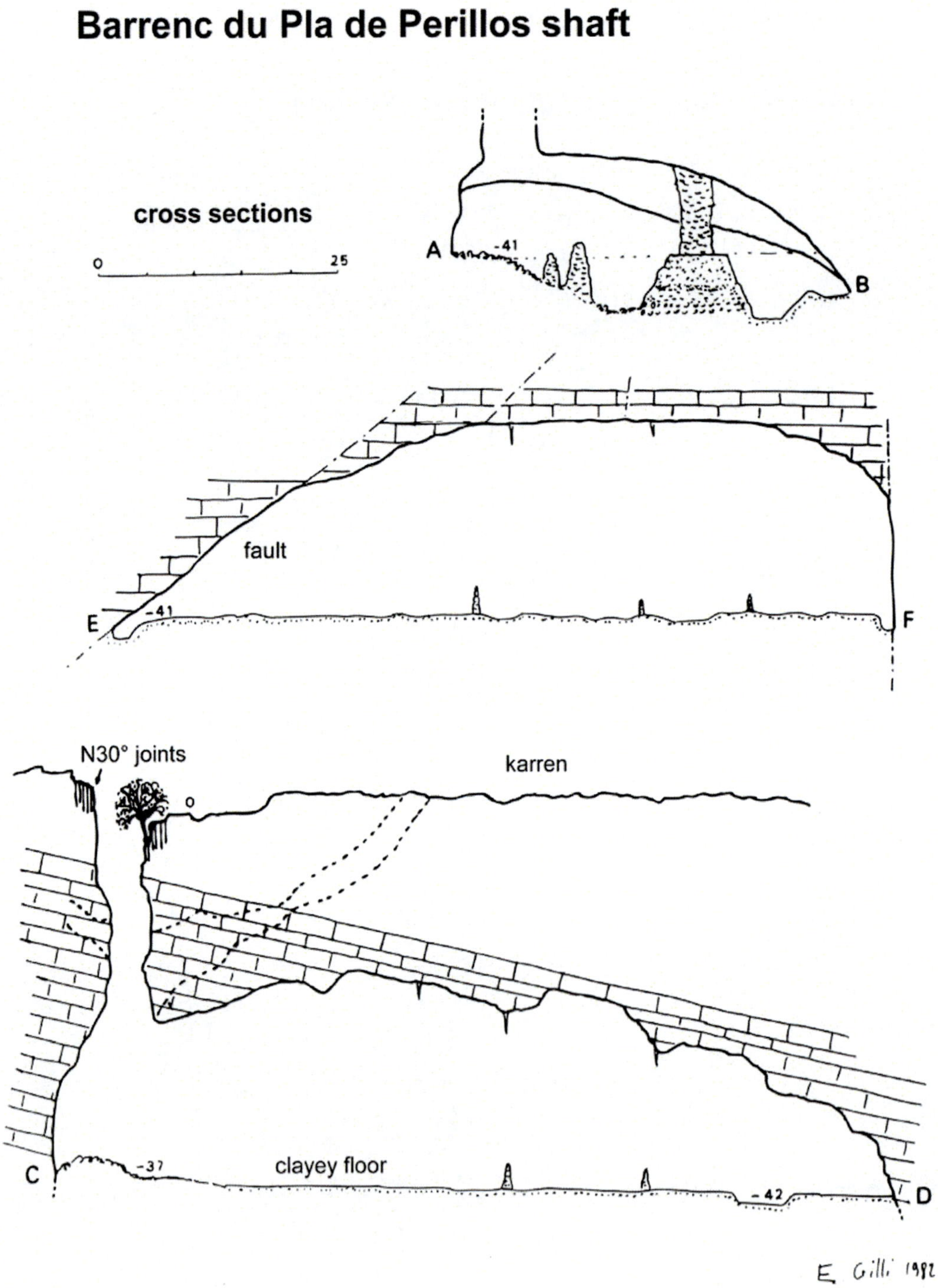

Fig. 2.128 Cross sections of Barenc du Pla de Périllos Shaft

2.19 Poudrey Shaft (Etalans, Doubs, France)

2.19.1 Description

See Figs. 2.129, 2.130, 2.131, 2.132, 2.133, 2.134, 2.135, 2.136, 2.137, 2.138.

This cave, certainly known for a long time by the locals, was explored in 1899 by the French geologist E. Fournier. It is now a tourist cave. It is located 20 km from Besançon on the N57 road to Lausanne. The entrance is at the edge of the road, 2 km after l'Hôpital-du-Grosbois (coordinates: 47° 10' 1.06" N; 6° 14' 44.21" E).

The Poudrey chasm is an oval room 130 m long by 100 m wide. The entrance is a pitch 27 m deep whose upper part is very flared and forms a sinkhole clearly visible in the landscape. The shaft leads to the top of a scree slope close to the western wall of the chamber, east of a set of large blocks of several meters, the highest of which is arranged into a terrace for tourists. This point is placed at the level of the ceiling. This one is remarkable. It is a slab at the base of a limestone bench with a span of 90 m.

The floor is sloping 20° to the east. The lowest point of the cavity is a collapse funnel. A second small funnel collects the water flowing from a small lake fed by a trickle of water arriving from an upper small conduit at the north of the chamber. Along the walls, an embankment reaches the foot of the vault, which is sometimes very calcified, especially in the southeast part.

2.19.2 Geology and Hydrogeology

The chasm opens on a hilly plateau with many sinkholes some of which exceed 20 m deep (Combe la Grise, Bois Brothel, Combe au Pussenard). The topography is softened by the presence of thick soil, cultivated or forested. The thickness of the limestone above the hall is only 25 m. The carbonate series is horizontal which straightened to the

Fig. 2.129 Chamber of Poudrey Shaft

Fig. 2.130 Flat ceiling of Poudrey Shaft

Fig. 2.131 Scree slope and marly levels of Poudrey Shaft

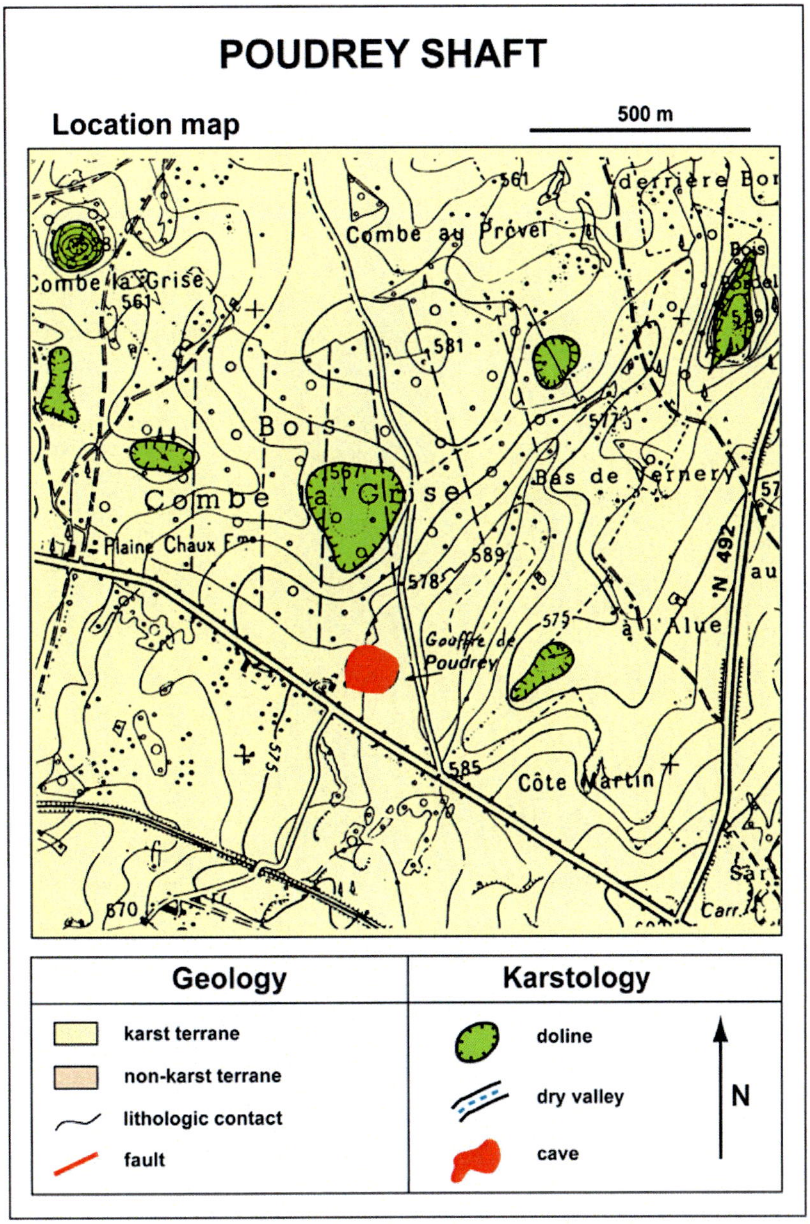

Fig. 2.132 Geomorphological map of Poudrey Shaft

Fig. 2.133 Marly wall acquiring an arched form

northwest and is framed in the northwest and the southeast by vertical N45° faults.

From the surface to the bottom, the formations are as follows:

- 30 m of fine fossiliferous white limestone, forming a compact mass dated Upper Sequanian;
- 10 m of red and beige massive limestone;
- few meters of marly limestone;
- fossil-bearing blue marls with intercalations of limestone. This set is dated from the Middle Sequanian (Oxfordian) (Fig. 2.133); its thickness is around 30 m. The foot of the vaulted part of the ceiling is dug in the marls (see AB and CD sections) (Figs. 2.138, 2.139).

It is not possible to see the nature of the rock in the draw funnels; however, the Lower Sequanian and Rauracian limestones are presumably very close.

The layers are horizontal. One can note the presence of some fractures:

- A set of N5° vertical diaclases. It was at their expense that the entrance shaft opened.
- Some diaclases of variable directions, which cut the ceiling without affecting its stability.

There are several parts of cave filling:

- an earthy cryoclastic scree with fine elements which corresponds to the withdrawal of the entrance sinkhole;
- a decimetric block scree with an earthy matrix that occupies most of the cave; it covers metric blocks;
- a scree slope located at the foot of the arched ceiling in the southern part of the room. This slope partially masks the blue marls and is formed of their alteration products as well as of elements coming from the ceiling (40 cm large blocks). The top of the embankment is cluttered with decompression scales;
- clusters of large blocks on the northwestern and north-eastern parts of the chamber; these limestone blocks are tilted due to the scouring of the marls on which they lay;
- the bottom part of the room is hollowed out by a funnel which reveals the presence of lower karst conduits that make possible the racking of the marls. This funnel deepened in 1980 during a big flood;
- speleothems are present in different parts. On the ceiling, there are some stalactites aligned on the joints. Large masses of flowstone cover the marly slope at the foot of the vault, especially in the southeastern area of the chamber. The water drips from the limestone located above the marls. Around the funnel, the flowstone is disorganized and tilted which confirms the withdrawal of materials by lower groundwater flows (Fig. 2.134).

To the north of the room, an outlet located high in the wall and a trickle of water from the ceiling feed a small artificial lake which is swallowed in a 2-m-deep hole. A dye test proved a relation with the Bremen spring a few kilometers to the south. There are certainly more important water circulations below the room, allowing the racking of the marls.

Fig. 2.134 Tilted stalagmites on the edge of a racking funnel

2.19.3 Genesis and Stability

The following mechanism is assumed:

Underground circulation took place at the base of the limestones at the top of the most marly levels. In a second time, the scouring of the marls under the Sequanian limestone slab generated the chamber. Enlargement was followed by a tilting of the walls when the overhangs became too large. These walls have gradually stabilized forming vaults. At the top, the vault rose up to reach the slab where the collapses were stopped, like in Champclos (Chap. 10) or in Mas d'Azil (Chap. 14) caves. The access shaft corresponds to the disorganization and gradual collapse of a dense diaclase zone. The bottom part of the chamber, at the racking funnel, is still active. Collapses took place during the 1910 floods. The foot of the vault in the eastern half of the room is affected by important phenomena of flaking in the marls (Fig. 2.135). At the bottom of the entrance staircase, the large tilted blocks correspond to the effect of the racking on the walls.

Fig. 2.135 Flaking of marls at the foot of Poudrey Shaft vault

Fig. 2.136 3D sketch of Poudrey Shaft

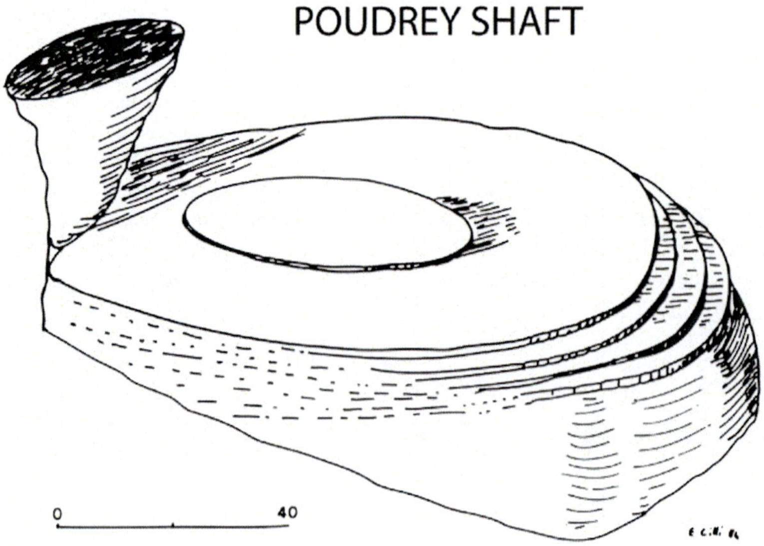

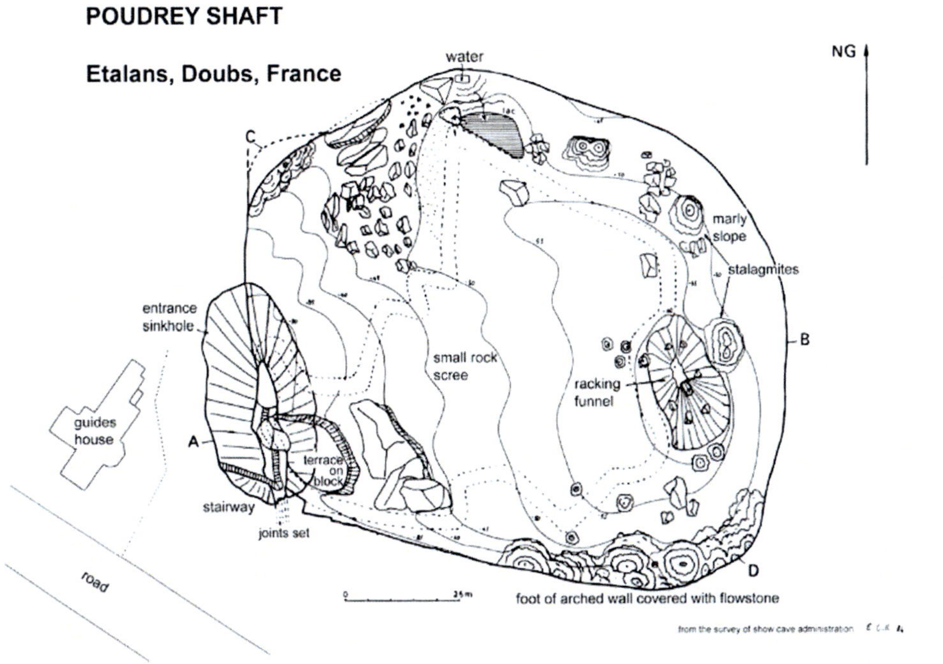

Fig. 2.137 Map of Poudrey Chamber

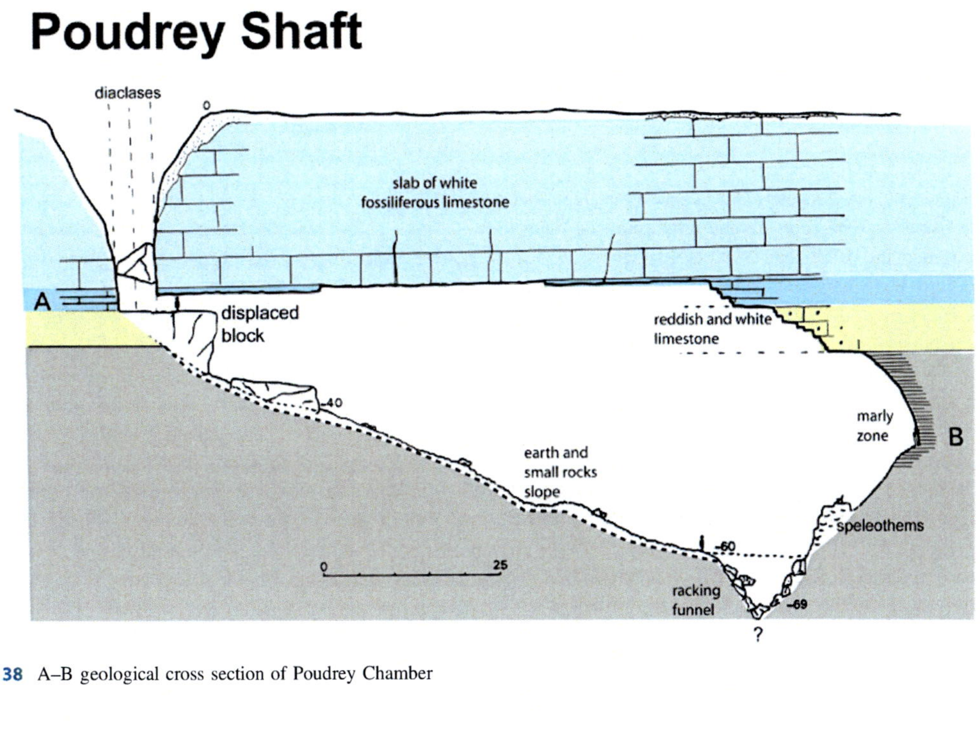

Fig. 2.138 A–B geological cross section of Poudrey Chamber

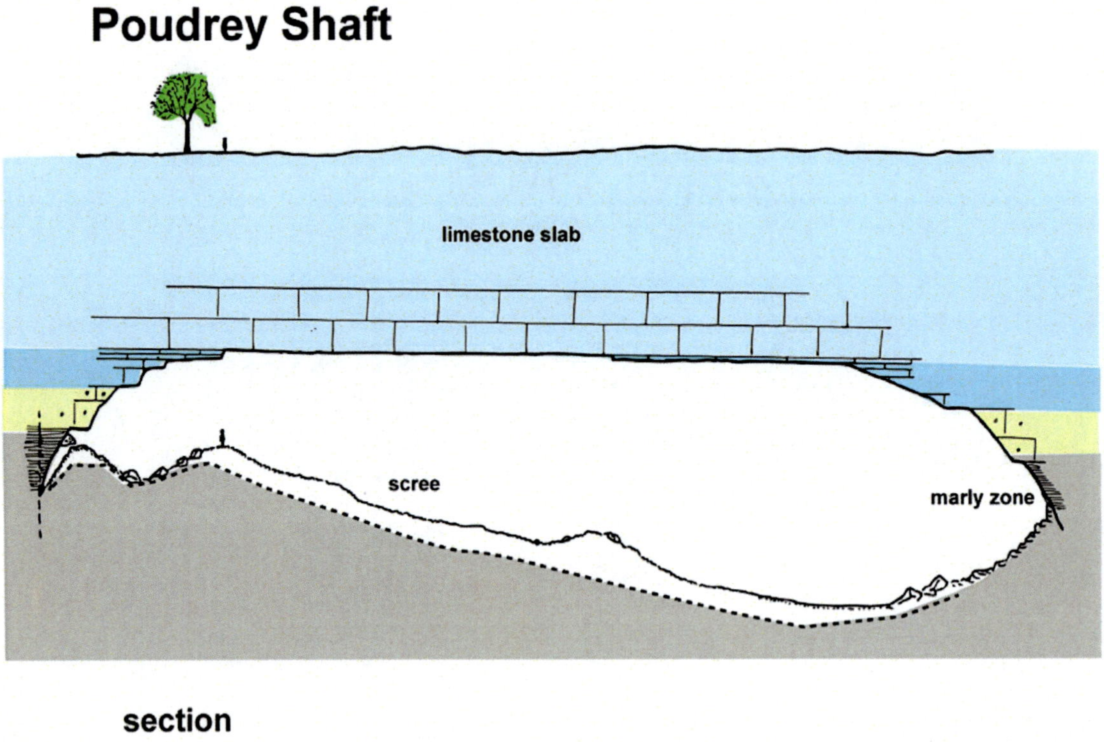

Fig. 2.139 C–D geological cross section of Poudrey Chamber

2.20 Pradières (Bédeilhac, Ariège, France)

2.20.1 Description

See Figs. 2.140, 2.141, 2.142, 2.143.

The Pradières Cave is located 150 m above the Bédeilhac Cave (see Chap. 3). The access is the same until the quarry which precedes the Bédeilhac visitors' car park. A few meters before the quarry, a footpath climbs up to the southeast and goes to the foot of the cliffs. From there, head west, then pass a flat and follow the cliffs down to the southeast. The entrance is well visible. It is located 200 m from the flat, at the foot of the cliff (coordinates: 42° 52' 7.17" N; 1° 34' 19.49" E).

The Pradières Cave is a giant gallery whose floor is covered with a scree. Its width reaches 65 m in places. It begins with a porch whose floor is powdery. The ceiling is vaulted, regular and 10 m high. The floor then becomes chaotic and, in some places, is covered with flowstone. The main gallery bends to the south; to the north a small gallery allows to cross the Sédour Peak to the north. It exits at the foot of the cliffs not far from the access path. To the south, the gallery continues to a square chamber whose southern corner is very calcified. Behind the stalagmites, a passage leads to an exit at the foot of the cliffs west of the Sedour Peak. The room is therefore located only a few meters from the slope.

2.20.2 Geology and Hydrogeology

The regional context and the lithology are the same than Bédeilhac Cave (Sect. 2.3).

The thickness of limestone above the gallery varies from 100 m.a.s.l. above the gallery bend to 20 m above the terminal square chamber. The gallery is still very close to the slope from which it never deviates more than 100 m (Fig. 2.140). The rock is very massive. The layers strike at the Sedour peak about N120° with a dip of 50° to the northeast. Inside the cave, the stratification is indistinguishable. Throughout the length of the cave, many fractures of variable orientation form a tight network. The main directions that are observed are N30° and N130°.

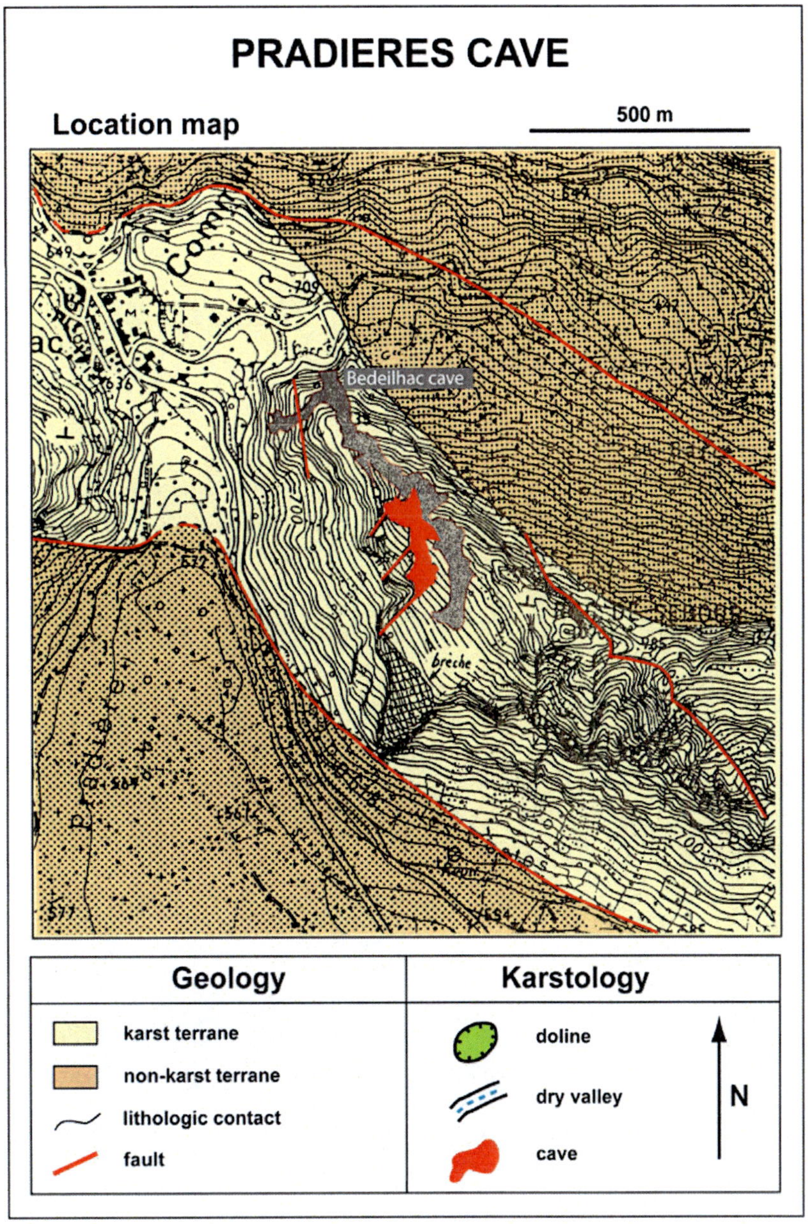

Fig. 2.140 Geomorphological map of Pradières and Bédeilhac Caves

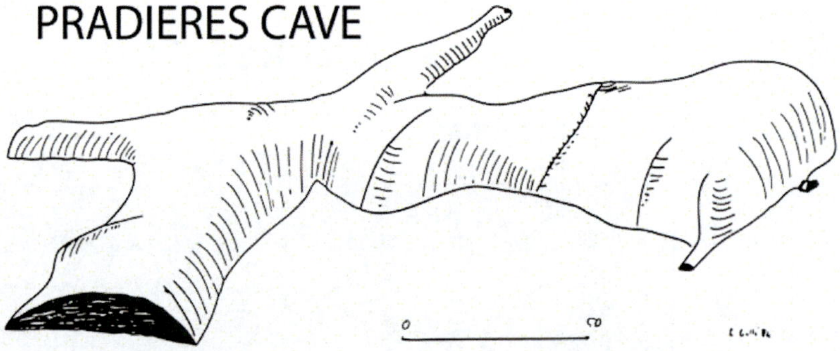

Fig. 2.141 3D sketch of Pradières Cave

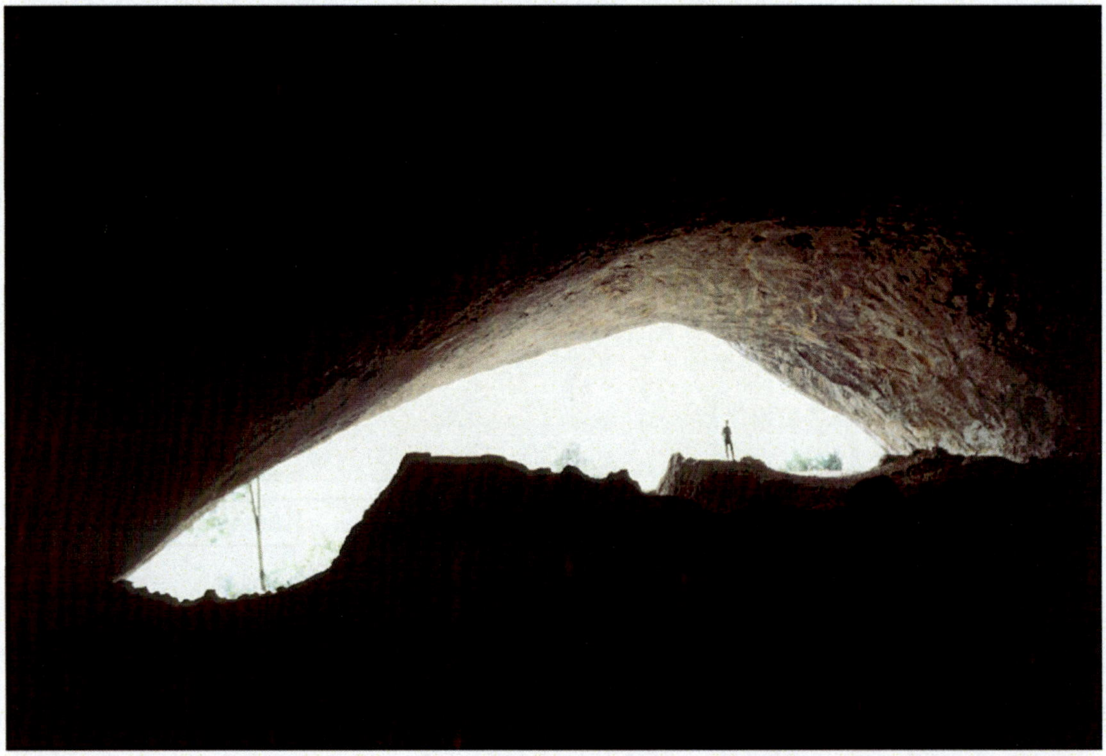

Fig. 2.142 Entrance porch of Pradières Cave

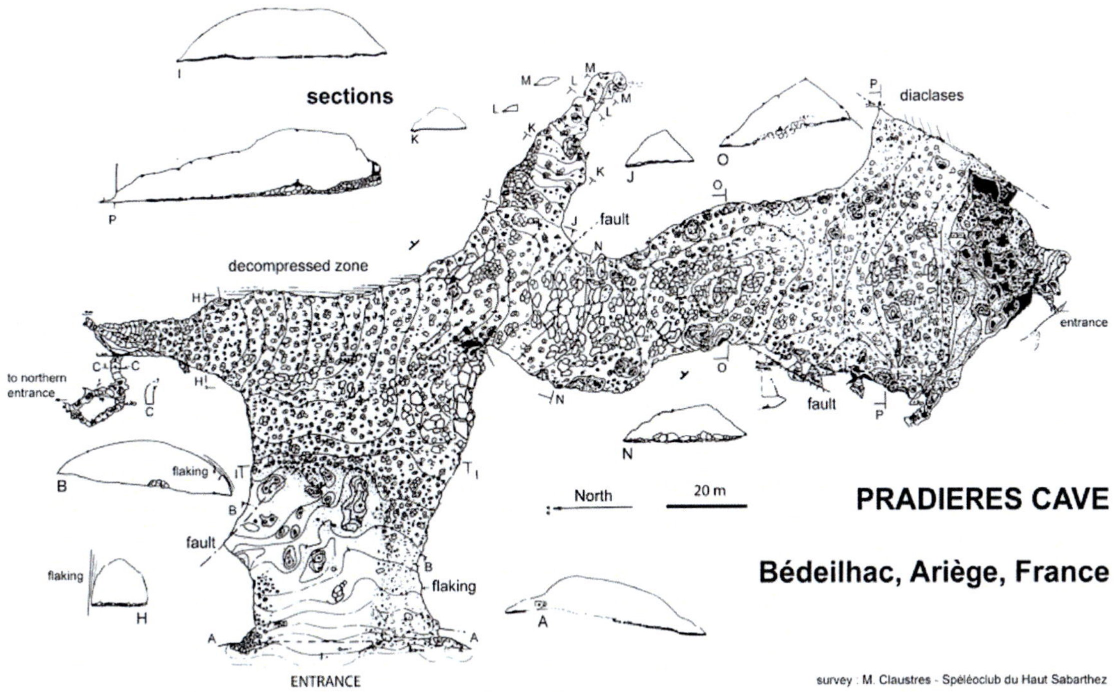

Fig. 2.143 Map and cross sections of Pradières Cave

They are of two kinds of cave filling:

- At the entrance of the cave, there is a thick powdery soil whose origin is unknown. It is likely a glacial deposit.
- In the rest of the cave, the soil is covered with angular blocks of variable size. Some are over 5 m long. They are, in places, covered with flowstone, especially in the southern corner of the terminal room. These blocks come from the ceiling and the walls. This scree is no more provided by rock collapses as the vaults have reached a very regular shape.

The cave is totally dry. It is a very old upper dry level of the Pradières-Bédeilhac Cave network.

2.20.3 Genesis and Stability

The Pradières Cave is a giant dissolution gallery shaped by the same mechanisms as Bédeilhac, Niaux, Lombrives or Sabart Caves.

The evolution by collapses of the ceiling continued after the abandonment of the gallery by the water flows. The variations of stresses in the massif, due to the deepening of the surface hydrographic network, caused successive collapses. The gallery today has reached a balance profile.

The cave is currently stabilized. One can, however, note a small zone very chipped at the bottom of the entrance gallery. It is perhaps still evolving. The entrance porch, of a very regular section, is visibly affected by cryoclasty.

2.21 Ravières Shaft (Orchamp-Vennes, Doubs, France)

2.21.1 Description

See Figs. 2.144, 2.145, 2.146, 2.147.

This shaft has been known for a long time. Around 1900, a masonry vault was built over the shaft entrance to prevent the population from throwing dead livestock and trash (Gigon and Monnin 1966). The access is by Orchamp-Vennes, northwest of Morteau. Drive 2 km on D41 toward the Ravieres hamlet. A few meters before the junction of the Saucet farm, cross the field located on the left of the road toward a barely visible small mound. This is the masonry vault covered by grass (coordinates: 47° 6' 37.33" N; 6° 30' 10.90" E). The entrance is located in the southern part of the structure.

An 18-m-deep pitch reaches the top of a scree cone located in the southern part of an oval-shaped chamber 65 m long and 55 m wide.

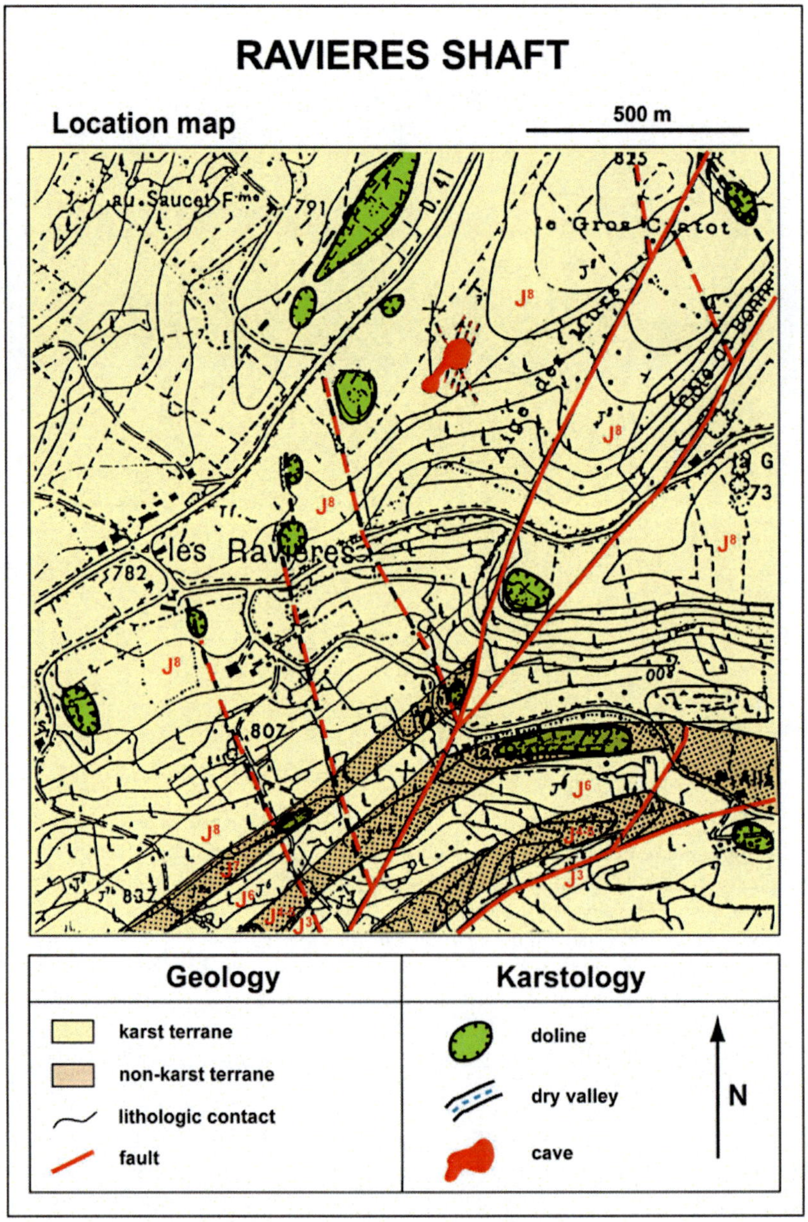

Fig. 2.144 Geomorphological map of Ravières Shaft

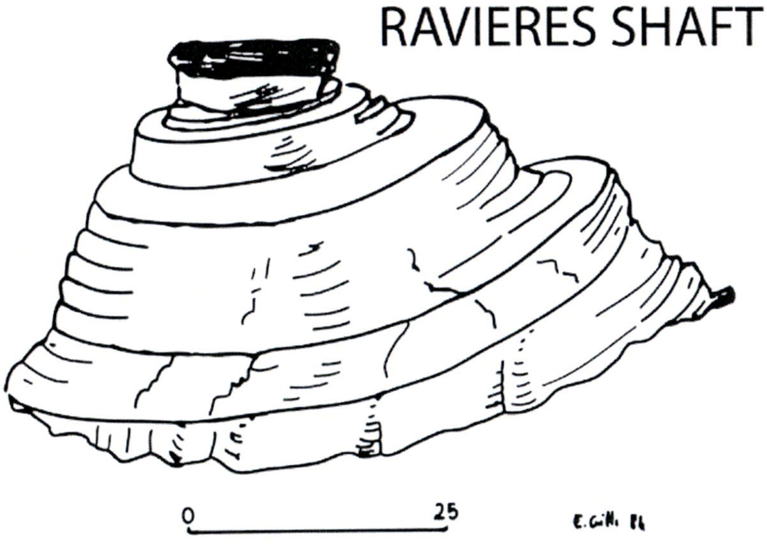

Fig. 2.145 3D sketch of Ravières Chamber

Fig. 2.146 The main chamber of Ravières Shaft

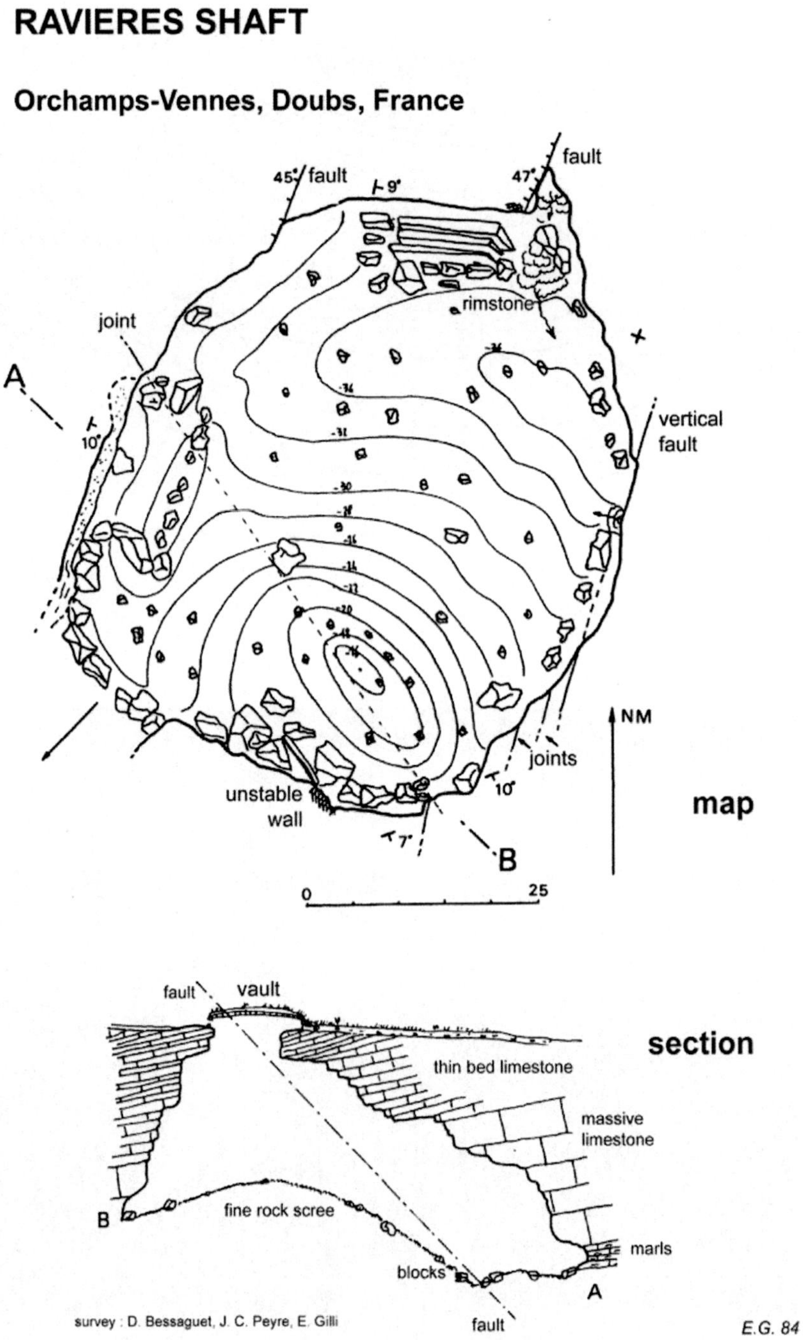

Fig. 2.147 Map and cross section of Ravières Shaft

The ceiling is inverse stair-shaped (corbelled vault) which is punctured in its high part by the entrance orifice which measures approximately 8 m long and 4 m wide. The hall is bordered by a blockfield. The center is occupied by a rather fine scree going down to the north to the low point at—6 m. A gallery whose departure is located southwest of the room leads to another room which is about 30 m in diameter.

North of the room, a trickle of water pouring from a recess of the wall gave birth to some rimstone dams. To the left of this recess, there are marl levels that have been scoured which provoked the collapse of the limestone beds that surmounted them. These marly levels are visible along the western wall.

2.21.2 Geology and Hydrogeology

The Ravières Shaft is 795 m.a.s.l. in a typical Jurassian landscape region. The limestone outcrops are highlighted by erosion and are separated by valleys dug in softer marly levels. South of the entrance to the cave, the vertical series accentuate this type of landscape. The entrance to the cave is located on the southwest flank of a hill that borders a small elongated plain. The plain corresponds to the center of a syncline and is devoid of surface water; it is punctured by many sinkholes aligned on N160° cracks. The shaft opens on the northwestern flank of a syncline whose axis direction is N50°. It is cut by an N20° vertical fault located 250 m from the cavity.

From the entrance of the shaft to the lowest point of the room, the series are

- 10–15 m of sublithographic yellow limestone stratified into small cracked beds; rock collapses are frequent;
- about 20 m of sublithographic yellow limestone in large metric beds that form a very massive unit;
- alternations of green marl and small limestone beds whose thickness is poorly determined.

These carbonate series belong to the Kimmeridgian. The Sequanian (Lower Kimmeridgian-Upper Oxfordian) limestone is probably very close. Same sequences are observed in the second room. The strikes are N25° to N70° with dips of 0–10° to the SE.

The room is very fractured and is crossed by three N20° faults. The easternmost is vertical and drives water circulations in the ceiling. The other ones are dipping 45° to the west. This direction is parallel to that of the main fault that cuts the syncline, southeast of the cave. A fourth vertical fault provoked the opening of the entrance shaft; its direction is N140°. Two diaclase sets parallel to both previous directions are clearly visible in the cave.

The cave filling is mixed. At the vertical of the entrance pitch, there is a scree of rather fine grain size at the top. It becomes coarser at its base and contains a few metric blocks. It is bordered all around the room by large blocks of metric size and by limestone slabs at the foot of the north wall. These slabs come from the collapse of the limestone layers caused by the scouring of the underlying marl at the foot of the wall.

The large blocks come partly from the breaking of these beds and partly from the collapse of the ceiling. The small blocks come from the profile adjustment of the vault; some may also come from outside. There are two small water inlets aligned on the two faults of the eastern part of the chamber. The water then infiltrates between the blocks at the lowest point of the room.

2.21.3 Genesis and Stability

The presence of the marly levels played an important role in the digging and in the size of the chamber. The water circulation caused the widening of the gallery by mechanical erosion of the marls. When the limestone overhangs became too large, the vault collapsed. This digging mode involves the existence of karst drains under the room. This lower cave network is probably dug in the Sequanian (Lower Kimmeridgian) limestones.

The walls of the entrance pitch are unstable; the limestone is very fractured and the blocks which collapse from this part feed the scree. The large limestone beds of the lower half of the vault appear resistant and stable. In the southern part of the room, near the entrance pitch, a very tight network of joints forms an instability zone.

2.22 Réveillon Porch (Alvignac, Lot, France)

2.22.1 Description

See Figs. 2.148, 2.149, 2.150, 2.151, 2.152, 2.153, 2.154, 2.155.

From Gramat, take D840 toward Martel village and then D673 to Padirac. A small road goes on the left 450 meters after the crossroad, in front of Réveillon hamlet. The cave is located below the first bend at the foot of an escarpment. A well-marked trail leads to the entrance (coordinates: 44° 49' 30.98" N; 1° 39' 53.57" E).

The cave entrance is great. It is a 35-m-wide porch located at the bottom of a limestone circus in which cascades the river that gave birth to the cavity. The porch is lengthened; the river runs on the rock between two earthy scree slopes bordering the walls. At the entrance, the flat ceiling is 40 m from the floor and then lowers to 7 m at the bottom of the porch. In the center of the ceiling, there is a collapse bell.

2.22.2 Geology and Hydrogeology

The cave is located on the Causse-de-Gramat Plateau on the edge of the Alvignac Dome where the Liassic marls outcrop. Surface water sinks at the contact between the marls and the Bajocian and Aalenian limestones. An escarpment underlines this geological contact (Fig. 2.150).

Above the porch, a dry valley, well visible in the landscape, is the old riverbed. The groundwater was initially flowing at the contact between the Liassic marls and the Bajocian limestone. The presence of dry galleries located on the southern wall of the porch at the elevation of this

Fig. 2.148 Inside view of Réveillon Porch

Fig. 2.149 Outside view of Réveillon Porch

stratigraphic contact testifies to this phase of sinking. The bed then deepened to give the current form (Renault 1968).

The cavity is located west of the Alvignac Dome in monoclinal series dipping to the west. From the top of the porch to the bed of the river, the series are

- 30 m of massive white limestone dated Bajocian.

Below is the Aalenian:

- 7 m of massive red sandy limestone;
- 15 m of black sandy limestone in small benches toward the base;
- 5 m of alternation of limestone beds and marly beds;
- the floor is made of small beds of black sandstone limestone.

The Toarcian marls are probably very close.

The strike is N20°, and the dip is 18° to the west. The porch is located between two faults:

- to the north, an N120° fault inclined 80° to the north; the wall of this fault forms the northern wall of the porch;
- to the south, an N90° fault with a dip of 80° to the south; the dry gallery at the contact between the Bajocian and the Aalenian is aligned on this fault line.

The underground riverbed is bordered by two embankments, the largest of which is located to the north. These slopes consist of small blocks in a clayey material. Locally, there are some clusters of large blocks, especially under the collapse bell of the ceiling, and at the foot of the north wall, in its middle part, in an area cut by the north fault.

The river flows on the limestone or on a thin bed of pebbles. The Réveillon Cave is the sink of the Salgue River that drains a depression in the Toarcian marls west of Alvignac. The water emerges at the Limon spring in the Dordogne Valley, 8 km west of the loss (Gèze and Cavaillé 1977). The cave includes a wet cave network as well as two dry levels that correspond to the gradual sinking of the waters down to the impermeable level formed by the Toarcian marls.

2.22.3 Genesis and Stability

The porch results of widening of the riverbed when the water passed through the marly levels at the base of the Aalenian. This enlargement created cantilevers in the limestone series of small beds which caused their collapse. The ceiling excavated upward until reaching the more compact limestone beds of the Upper Aalenian and the Bajocian.

Apart from a few rock scales falls, the vaulted ceiling is stable. At the foot of the walls, the Lower Aalenian marls present signs of instability near the N90° fault. Detached blocks are visible on the opposite wall near the north fault.

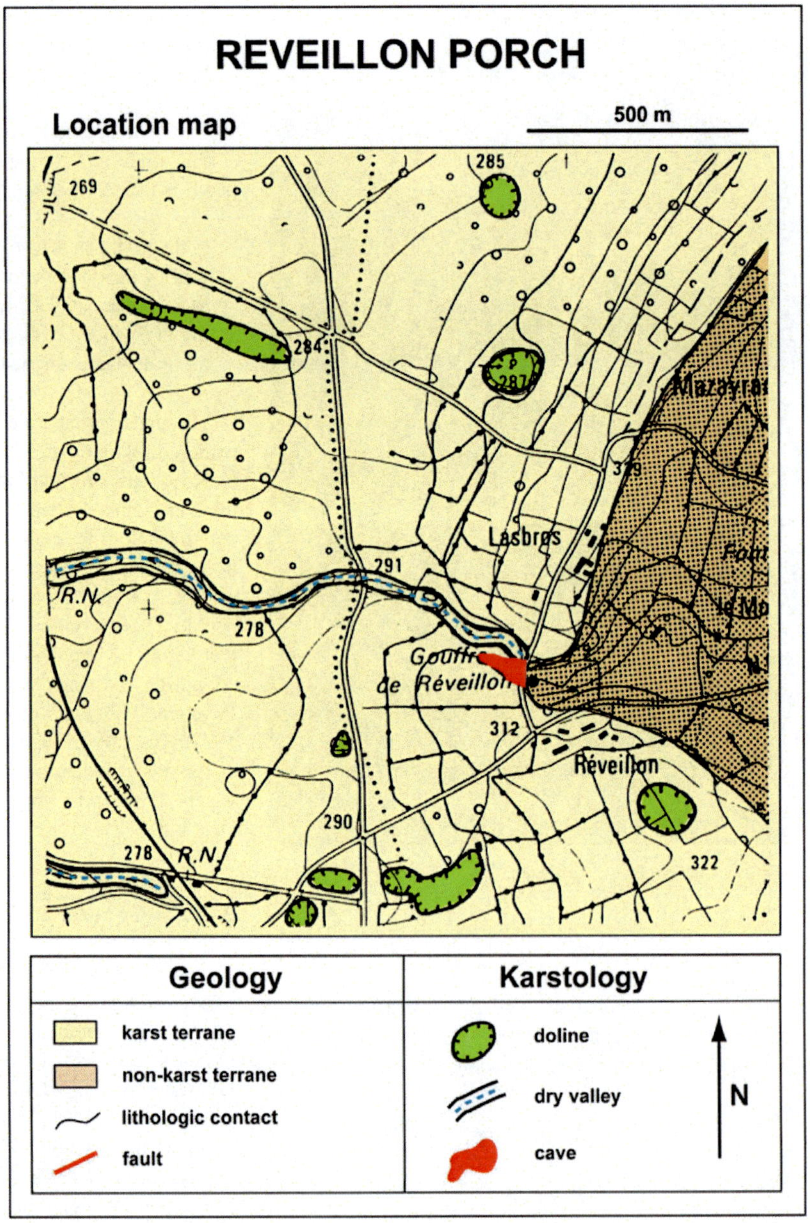

Fig. 2.150 Geomorphological map of Réveillon Porch

Fig. 2.151 Aalenian series of Réveillon Porch

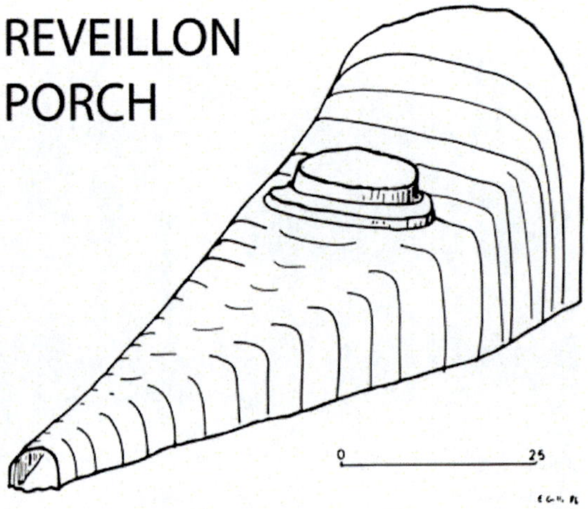

Fig. 2.152 3D sketch of Réveillon Porch

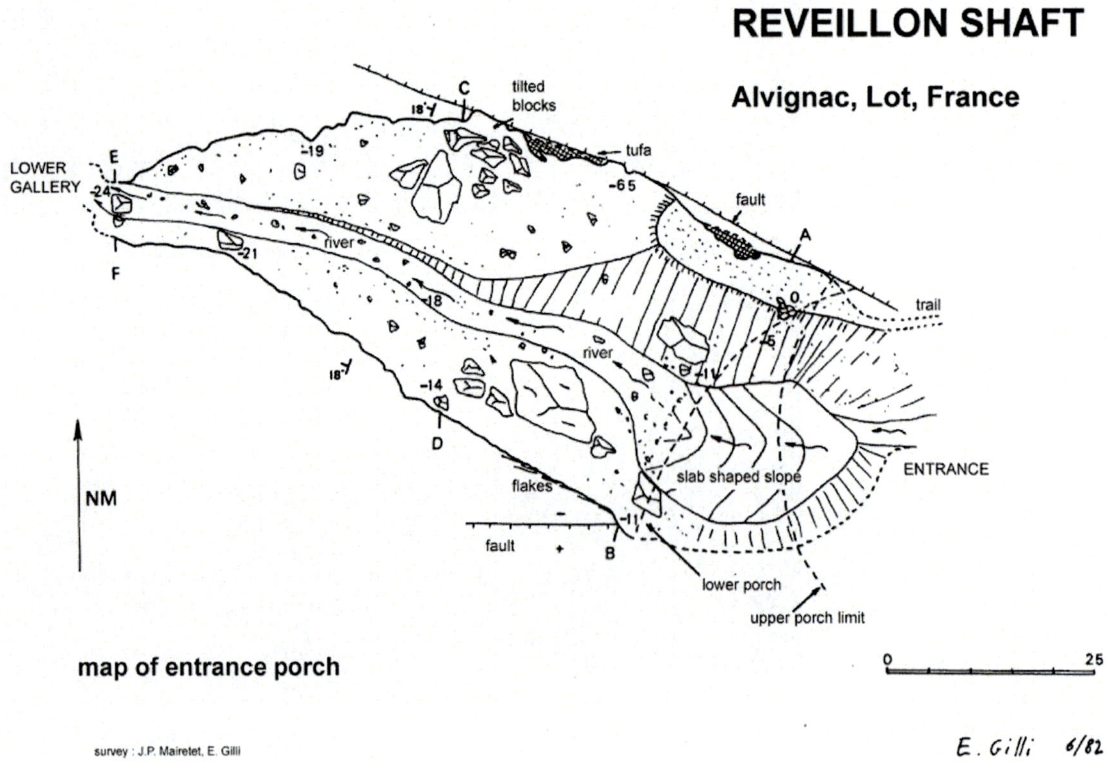

Fig. 2.153 Map of Réveillon Porch

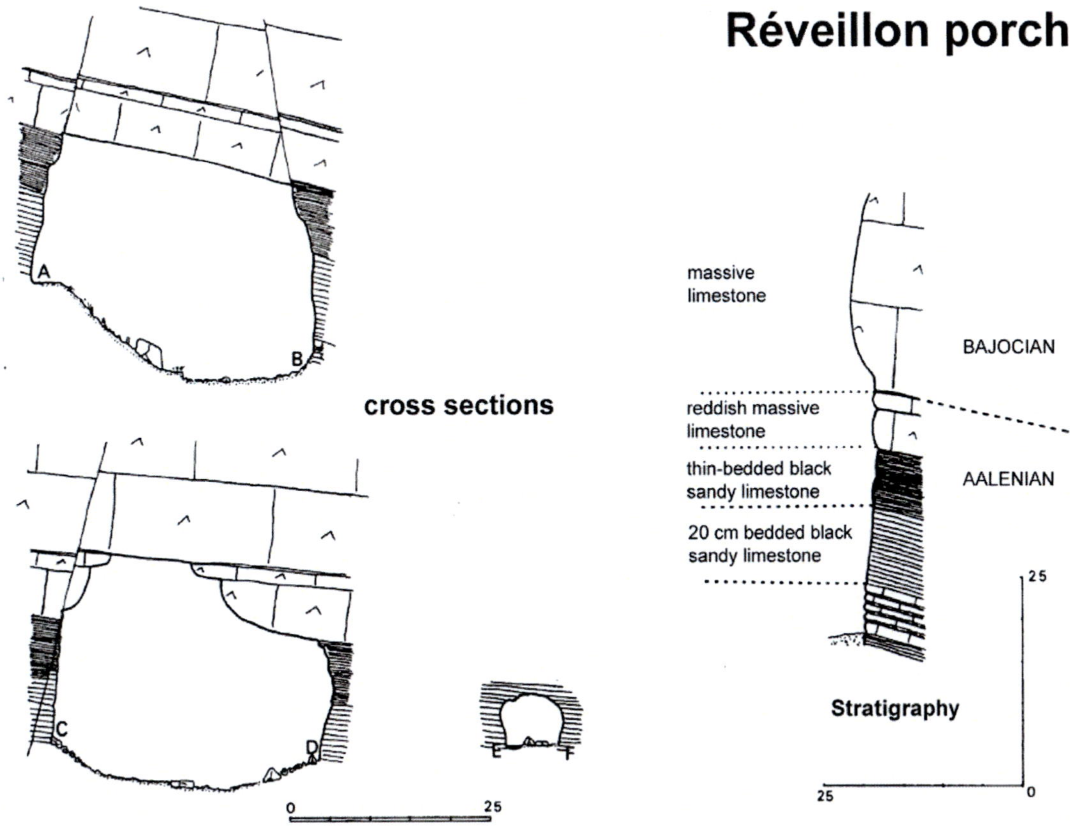

Fig. 2.154 Geological cross sections of Réveillon Porch

Fig. 2.155 Longitudinal cross section of Réveillon Porch

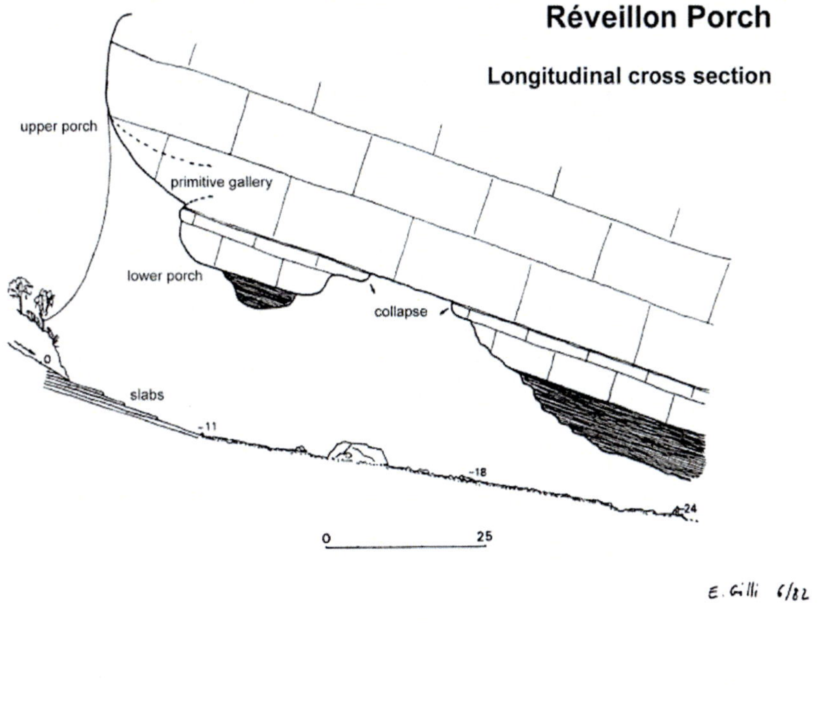

2.23 Riusec Cave (Portet-D'Aspet, Haute-Garonne, France)

2.23.1 Description

See Figs. 2.156, 2.157, 2.158, 2.159, 2.160, 2.161, 2.162, 2.163, 2.164, 2.165, 2.166, 2.167, 2.168, 2.169.

From St. Girons take to the west the D618 road toward St. Gaudens to the Portet-d'Aspet pass. From there, a wide and clearly visible path climbs to the west and crosses the forest of Portet-d'Aspet. After 3/4 of an hour walk, take on the right a small path little visible that climbs to the north, through a boxwood zone, to the Espugalon forest. This path goes down toward the valley of Rieusec and then goes up toward the cliffs located in the southwest of the Mail-de-l'Auech area. The cave is located on the cliffside about 1 h and a quarter walk from the Portet-d'Aspet pass. It is around 1300 m.a.s.l. The cave was studied by Trombe (1947) and Trombe et al. (1947).

The Riusec Cave is part of the Burtetch-Riussec network that extends on 3 km (Bigot 2004). The entrance porch gives access to a horizontal gallery that ends by a small jump. It overlooks a little chamber from which two galleries depart:

- to the east the first gallery leads to the Trombe Hall (130 m long and 90 m wide);
- to the north, the second gallery leads to the A Hall (125 m long and 95 m wide).

2.23.1.1 Trombe Hall (Fig. 2.157)

From the junction room, a chaotic gallery rises to the east, widening more and more until reaching a zone of huge blocks several meters wide. The hall starts here. It has roughly a semicircular shape; the floor of its northern part is a huge slope of schistous elements inclined 20–35° to the south. The southern part is occupied by a big blockfield that surrounds a gigantic block 60 m long. It is possible to walk along the rock at the foot of the vaults.

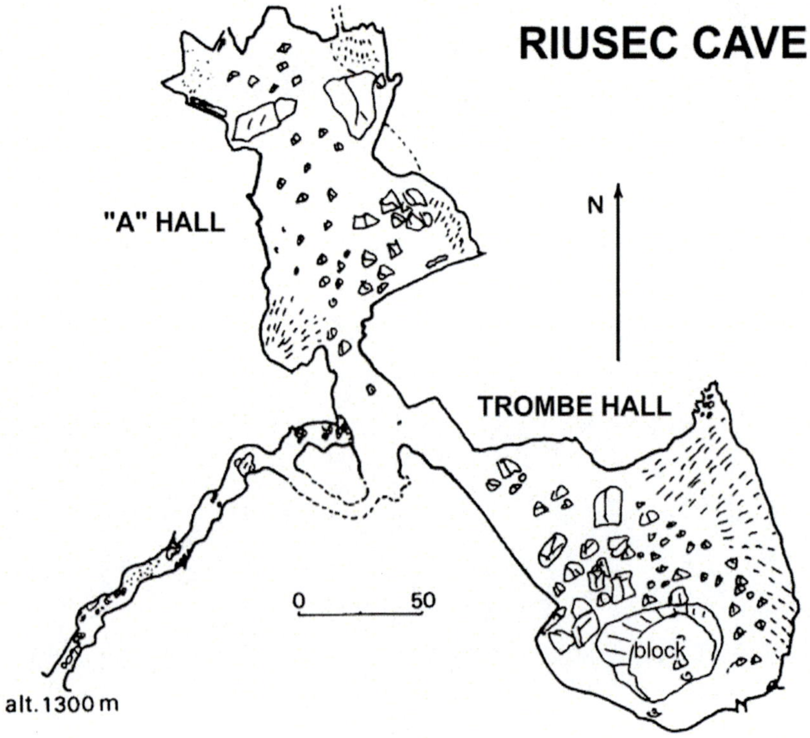

Fig. 2.156 Location of "A" and Trombe chambers in Riusec Cave

Fig. 2.157 Trombe Hall

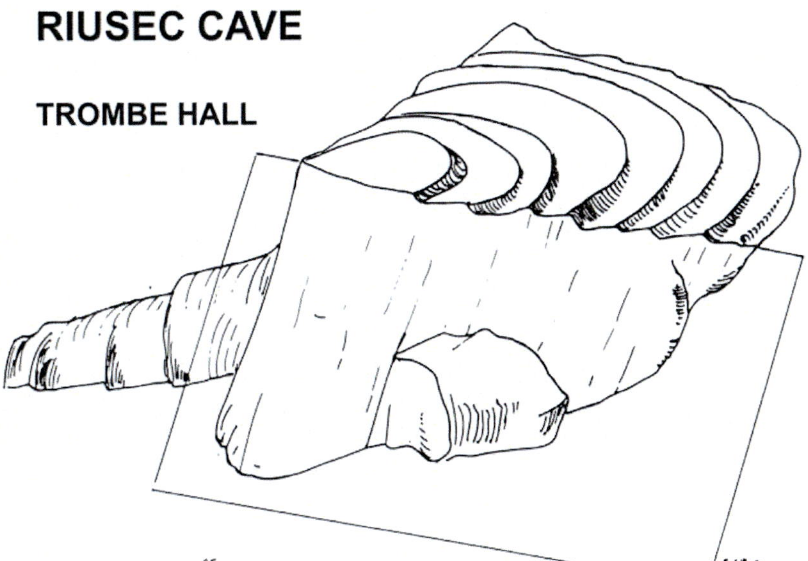

Fig. 2.158 3D sketch of Trombe Hall in Riusec Cave

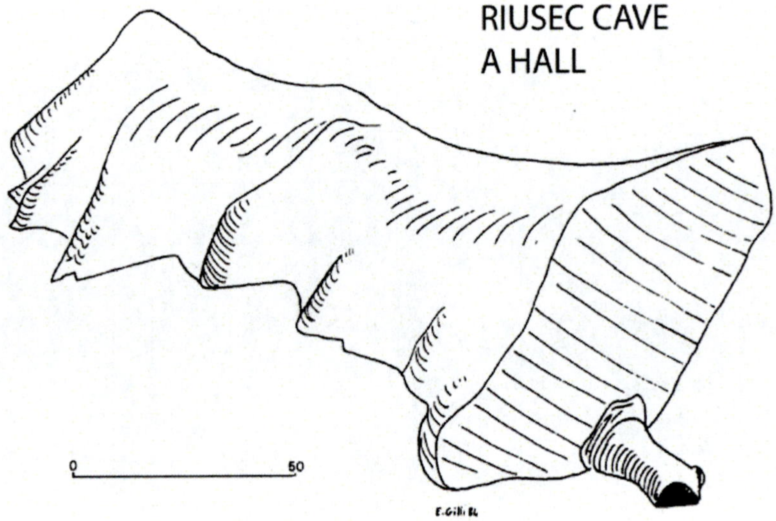

Fig. 2.159 3D sketch of “A” Hall in Riusec Cave

Fig. 2.160 “A” Hall. Note the fault wall in the background

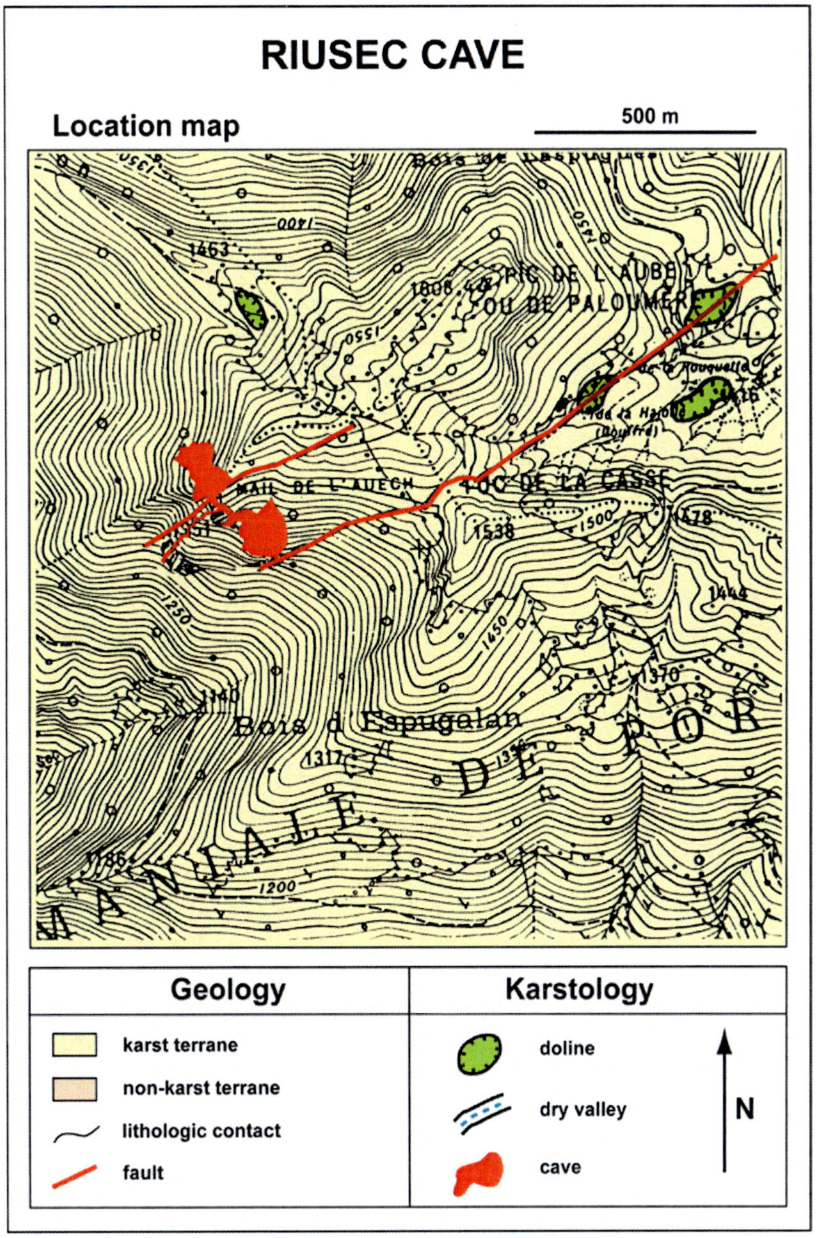

Fig. 2.161 Geomorphological map of Riusec Cave

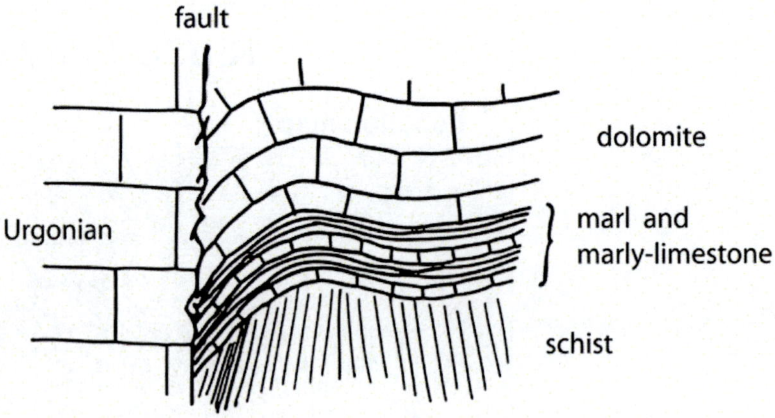

Fig. 2.162 Northern angle of Trombe Hall

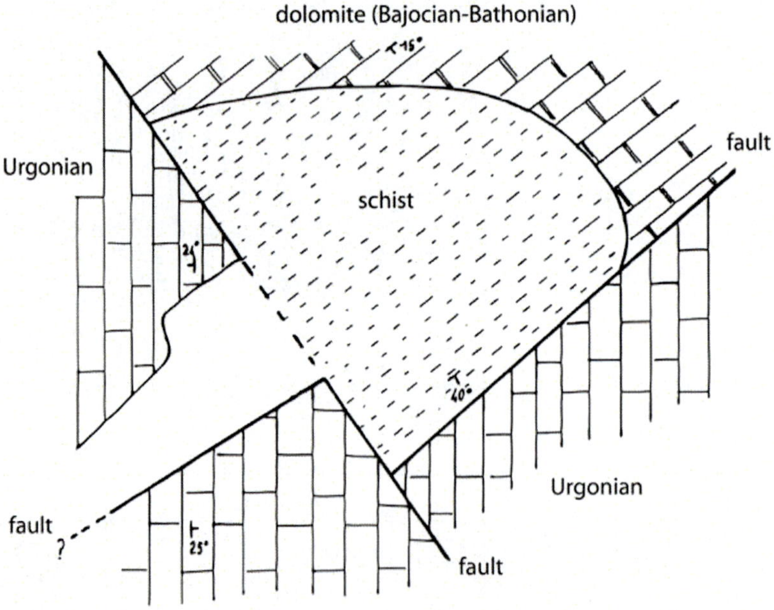

Fig. 2.163 Simplified geological sketch of Trombe Hall

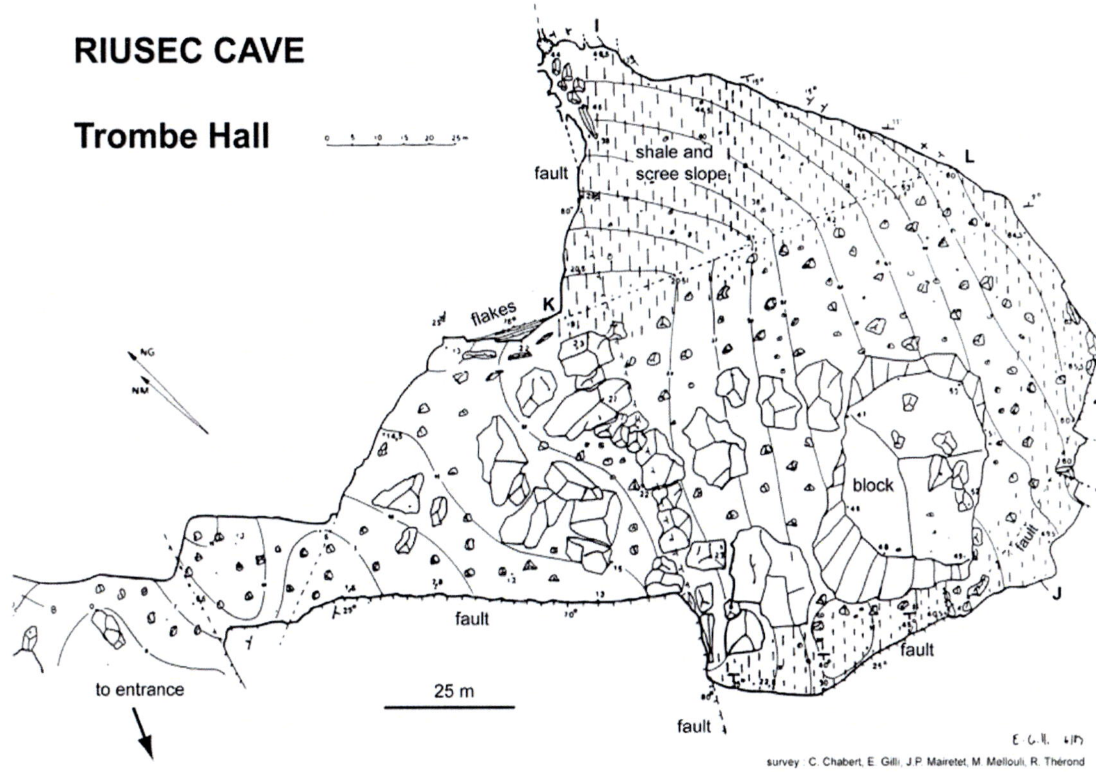

Fig. 2.164 Map of Trombe Hall in Riusec Cave

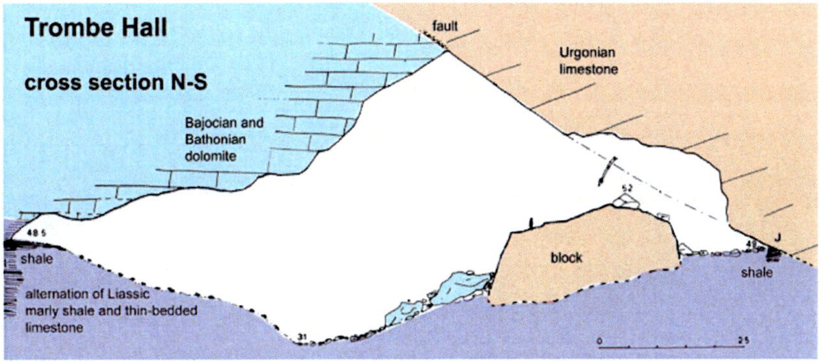

Fig. 2.165 N-S geological cross section of Trombe Hall

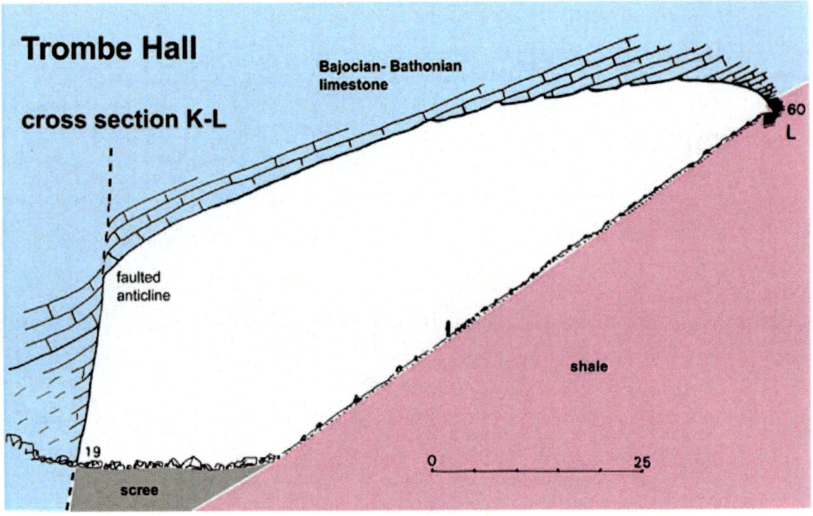

Fig. 2.166 K-L cross section of Trombe Hall

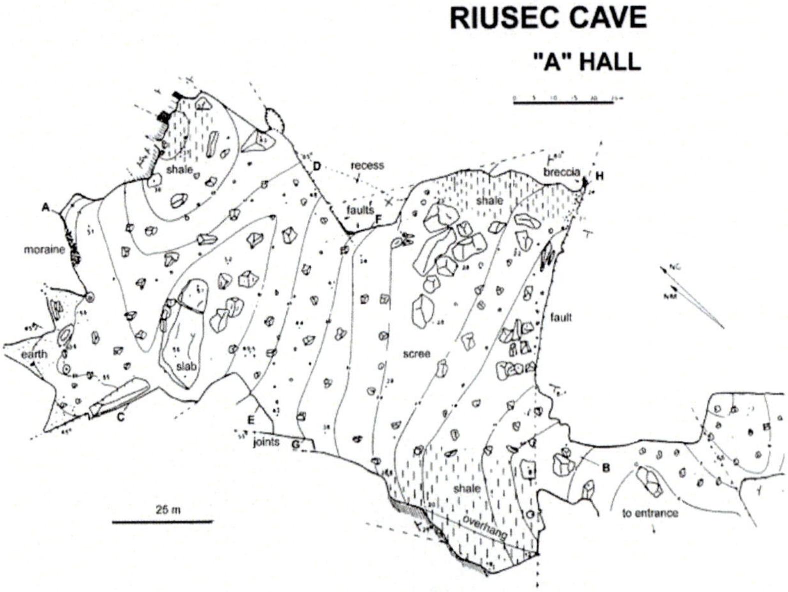

Fig. 2.167 Map of "A" Hall in Riusec Cave

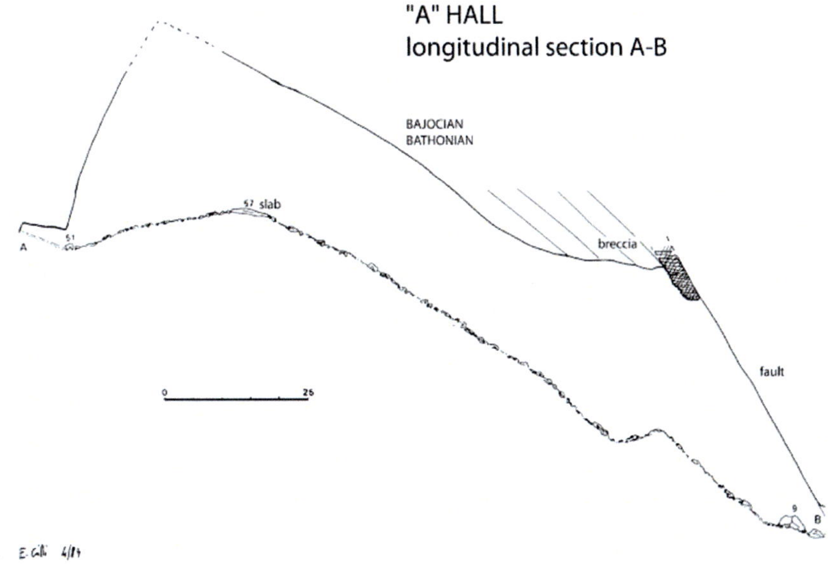

Fig. 2.168 Longitudinal section of A Hall in Riusec Cave

The ceiling is irregularly shaped. In its northern half, it is a plane, inclined 20° to the west. In the southern part, it is a large collapse bell. The highest point is about 40 m above the ground. The south wall of this bell is a fault plane inclined 30–40° to the south.

2.23.1.2 A Hall (Fig. 2.160)

From the junction room, a very short gallery gives access to the A Hall. The floor is a blockfield of big elements, at the top of which there is a 25-m-long block shaped as a slab. This scree goes down to the north. At its bottom, there is the entrance of the Rain Cave network which is part of the deep groundwater drain of the karst system.

In the north, south and east corners of the hall, there are schist scree slopes comparable to those visible in the Trombe Hall. The east corner presents some clay-earthy recesses; one of them (point A of section A–B, Fig. 2.168) is blocked by a morainic filling. The walls are more vertical than in the Trombe Hall and each mostly corresponds to the wall of a fracture, especially the southeast wall, above the entrance to the room. The highest point of the ceiling is located above the slab which is the summit of the scree. From this point, the ceiling dips to the south.

2.23.2 Geology and Hydrogeology

The cave is on the southwestern side of the Paloumère Peak (1608 m) at the foot of the Riusec Rock, 1300 above sea level. This rock is to the southwest of a crest called Mail-de-l'Auech under which are located both rooms (Fig. 2.161). The highest point of Trombe Hall floor is about 20 m from the edge of the cliff that limits to the south the Mail-de-l'Auech crest. The A Hall is about 80 m from the north side. The thickness of limestone above the chambers is about 100 m for A Hall and 80 meters for Trombe Hall.

Riusec Cave is located south of the Milhas Massif and its metamorphic border. It extends in a system of small thrust slices within the Jurassic and Cretaceous sedimentary series. The terrains are intensely folded and fractured. From top to bottom, the normal series includes the following facies (from the BRGM 1/80,000 geological map):

- light gray compact limestone with Urgonian facies, aged Aptian;
- blackish dolomites with rare intercalations of compact limestone, aged Bathonian and Bajocian;
- these dolomites pass gradually to the Lias formed of black schistose marl with intercalations of small decimetric beds of yellow marly limestone;
- the Lower Lias comprises gray or yellow compact limestone that frequently become brecchoid or cargneuliform at their base.

In the cave zone, the stratigraphic order of the various units has been disturbed by the folds and fractures that affect the entire massif.

The floor of the Trombe Hall is dug into the Liassic marls and marly-limestone (Fig. 2.163). This ensemble is overlayed in concordance by the Bajocian-Bathonian dolomites in the northern half of the hall. The lithologic contact can be observed all around the room. On the other hand, on the

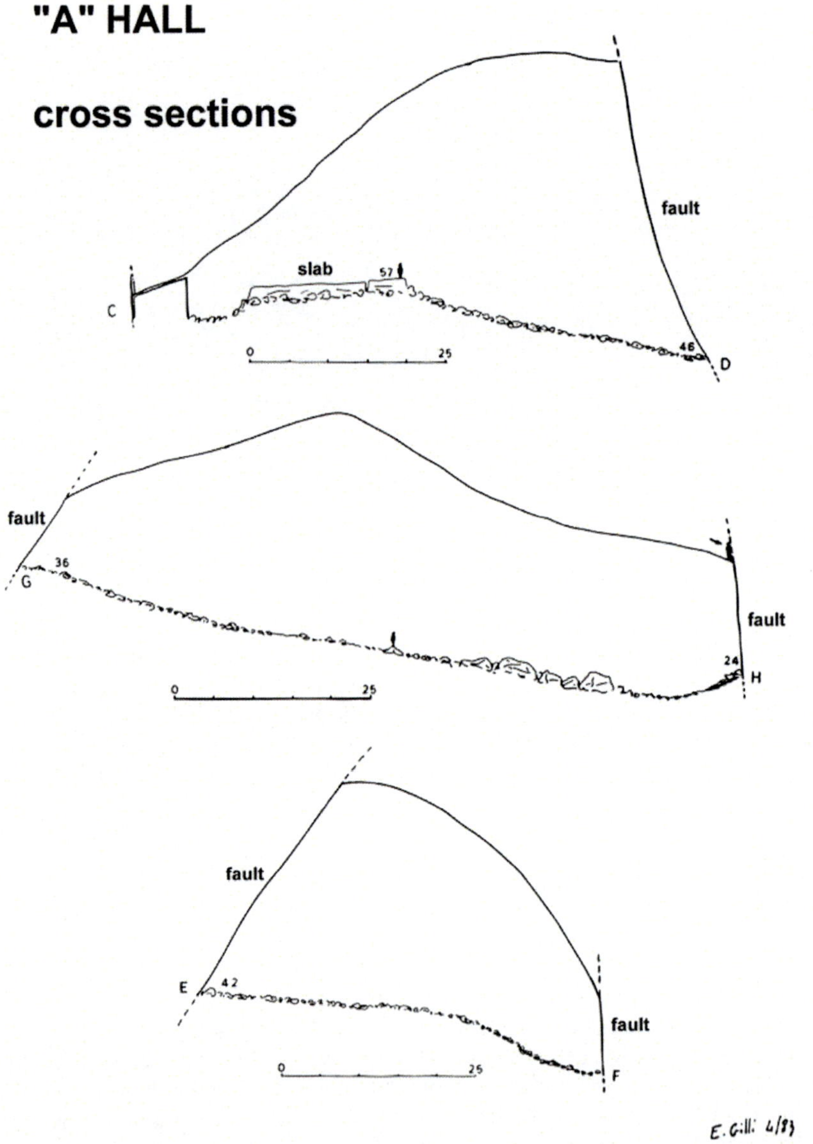

Fig. 2.169 Cross sections of A Hall in Riusec Cave

south, the schistose marls are surmounted by a limestone mass which is probably the Urgonian limestone brought there by the movement toward the south of a normal fault dipping 30–45° and whose wall forms the ceiling over the big block previously described.

The western limit of the chamber is the wall of an almost vertical large fault whose movement lowered the limestone of the western compartment until the schistose marl level. This limestone is presumably the Urgonian one. This abnormal contact is clearly visible in the northern corner of the room (48.5 m above the entrance).

The A Hall is more complex than the Trombe Hall. The rock is intensely fractured. In its lower part, at the southern and eastern corners, the Liassic schistose marl is present, surmounted in concordance by the Bajocian dolomites. This group is bordered on the south by a large fault almost vertical, putting the schistose marl in contact with a limestone mass which is undoubtedly the Urgonian. The top of the A Hall is dug in the Bajocian-Bathonian dolomites.

Near the entrance of the Rain Gallery, one can observe the transition from dolomite to schistose marl. In this part, the layers are straightened and folded; the tectonized zones indicate the near presence of faults. This part of the cave network requires special structural study because of the tectonic complexity. Given the intricacy of fracturing and tectonics, refer to Figs. 2.164, 2.165, 2.166, 2.167, 2.168 and 2.169 to visualize the essential features of the structure.

The most interesting observation is that fractures are numerous and form stable walls. Cave fillings are of two kinds:

- The slopes formed by schistose elements and small limestone blocks issued from the dismantling of the Liassic Marly limestone layers and partially covered with limestone blocks of variable size coming from the ceiling.
- The blockfields fed by the collapse of part of the ceiling and the walls.

In Trombe Hall, the 60-m-long block comes from the southern wall. Its tear-off niche is still visible. This block is not fractured; it has tilted gradually with this movement having been caused by the scouring of the underlying shales.

Concerning the hydrogeology, the wet part of the cave is the Rain Cave network through which an underground river flows. The water comes from the Paloumère Peak area. Several shafts make it possible to cross the massif by following this underground river. The discharge is around a few liters per second. The outlet of the water is at the Bleu spring, in Le Ger Valley, southwest of the cave (BRGM 2017).

2.23.3 Genesis and Stability

A same process seems to be the origin of the digging of both rooms. It is essentially the withdrawal of Liassic erodible materials (schistose marl and marly limestone). This racking was possible, thanks to the activity of a river previously located at the chamber level, during the digging stage and which one now finds at depth in the Rain Gallery.

The erosion of the marl has progressively provoked the overhanging of the Bajocian dolomites and the Urgonian limestones. Beyond a certain limit, the limestone masses have collapsed, especially in the areas situated between fault planes. This phenomenon still seems active.

Close to the contact between the Liassic schistose marl and the Bajocian dolomite, the rock is much disintegrated, and its stability is precarious. This instability is particularly perceptible in the Trombe Hall, in the north corner and also in the highest point, east of the room. It is the same for the A Hall in its southwest corner. All the schist slopes are at the limit of stability. Runoff waters dig small gullies. The west wall of A Hall, where the Bajocian dolomites are exposed, is very diaclastic, and the rocky massif is disorganized. On the other hand, the walls of the large faults which constitute some of the lateral sides of both rooms look particularly stable.

In the Riusec chambers, the digging seems still in progress but the current mechanism is slow. It is essentially linked to the action of the percolation water which circulates through the rooms and digs small gullies in the schistose marl debris.

2.24 Sabart Cave: Big Chamber and Renouveau Chamber (Sabart, Ariège, France)

2.24.1 Description

See Figs. 2.170, 2.171, 2.172, 2.173, 2.174, 2.175, 2.176, 2.177, 2.178, 2.179.

The Sabart Cave is part of the 14-km-long Niaux-Lombrives-Sabart Cave network which crosses right through the rock of Cap-de-Lesse south of Tarascon-sur-Ariège (Bigot 2004). The cave, well known in the region, has been visited since the nineteenth century (Minvielle 1977). A synthesis was published by Renault (1983).

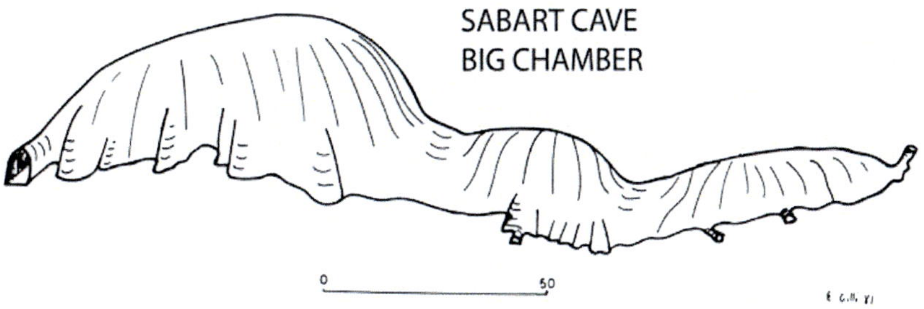

Fig. 2.170 3D sketch of Big Chamber in Sabart Cave

Fig. 2.171 The Big Chamber in Sabart Cave (photo Groupe Auscitain de spéléo)

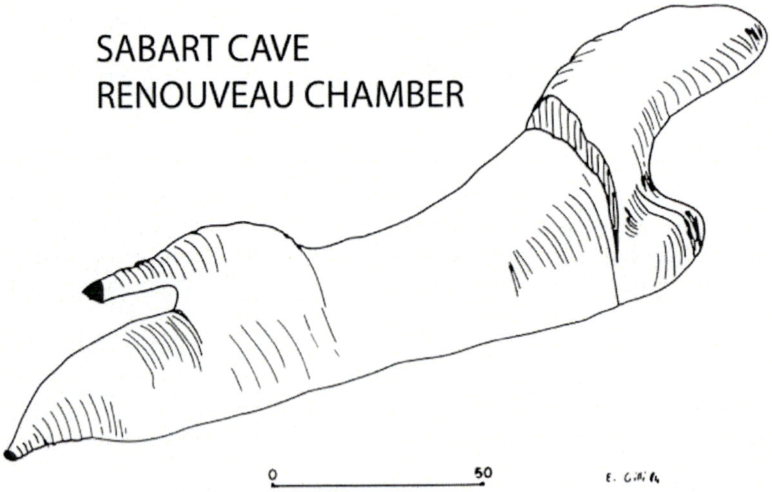

Fig. 2.172 3D sketch of Renouveau Chamber in Sabart Cave

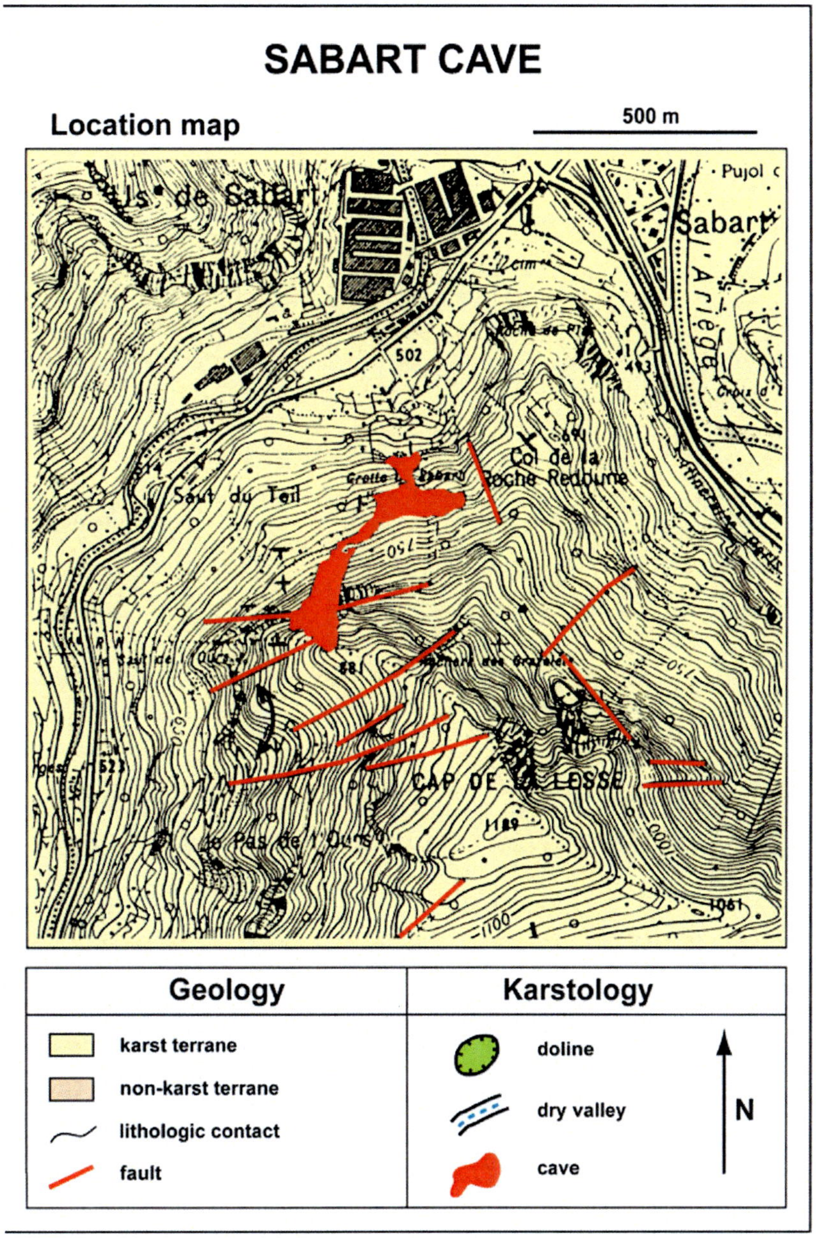

Fig. 2.173 Geomorphological map of Sabart Cave

The most obvious entrance is in a stone quarry, clearly visible from the D8 road from Tarascon to Vicdessos. A well-marked path climbs to the south, 400 m after the junction between the N20 and the D8, in front of the Sabart hydropower plant. At the back of the quarry, a very steep slope gives access to the Big Chamber through a barely visible entrance (coordinates: 42° 49' 56.24" N; 1°36' 0.85" E).

Access to the Renouveau Chamber is possible from the Big Chamber by taking, to the southeast the Grand Gallery, which narrows at its end until it becomes a tiny tunnel closed by a door whose key is kept by the Haut Sabarthez caving association. This door gives access to the room.

The Big Chamber is a giant gallery 190 m long and 90 m wide (Figs. 2.170, 2.171). There are two parts:

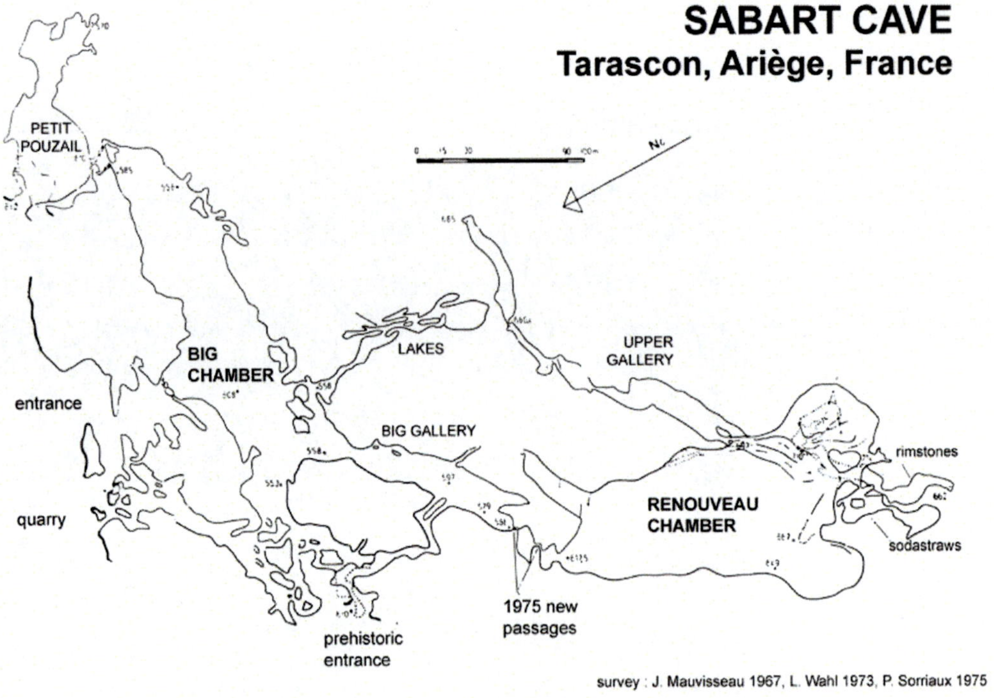

Fig. 2.174 General map of Sabart Cave

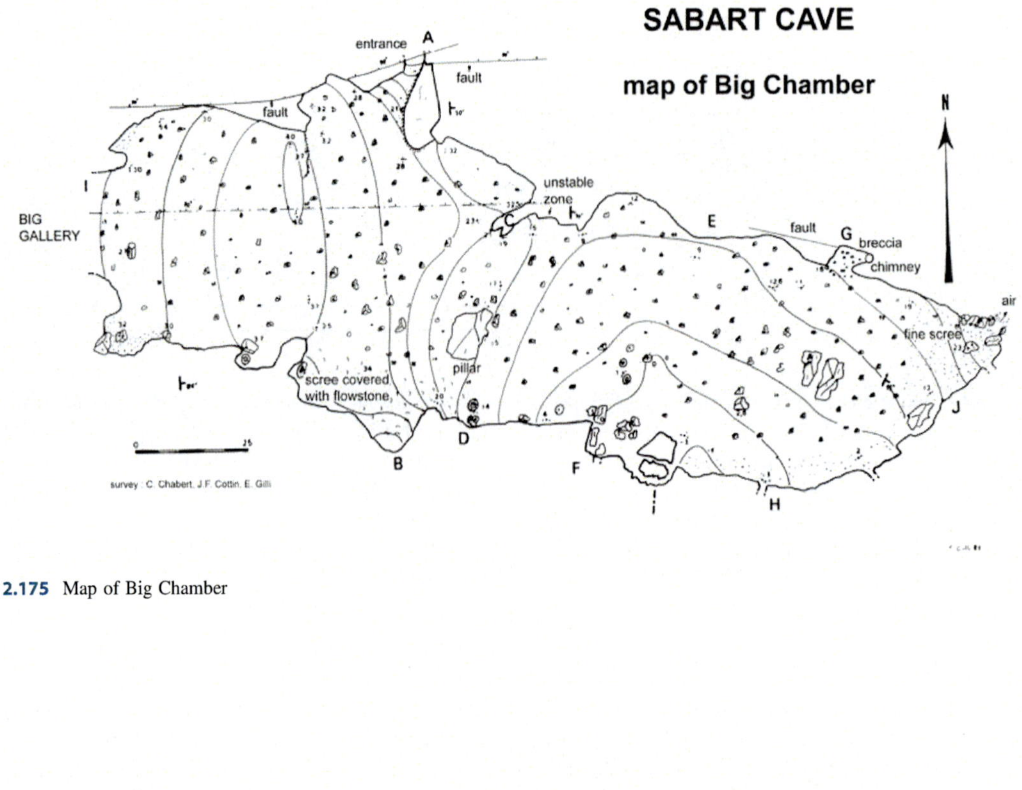

Fig. 2.175 Map of Big Chamber

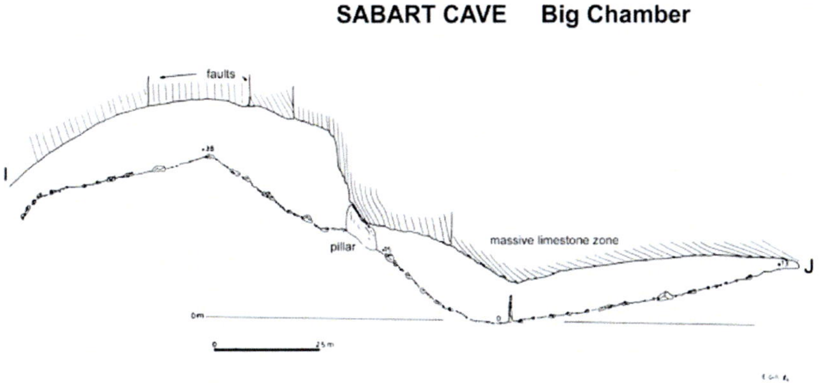

Fig. 2.176 Longitudinal section of Big Chamber

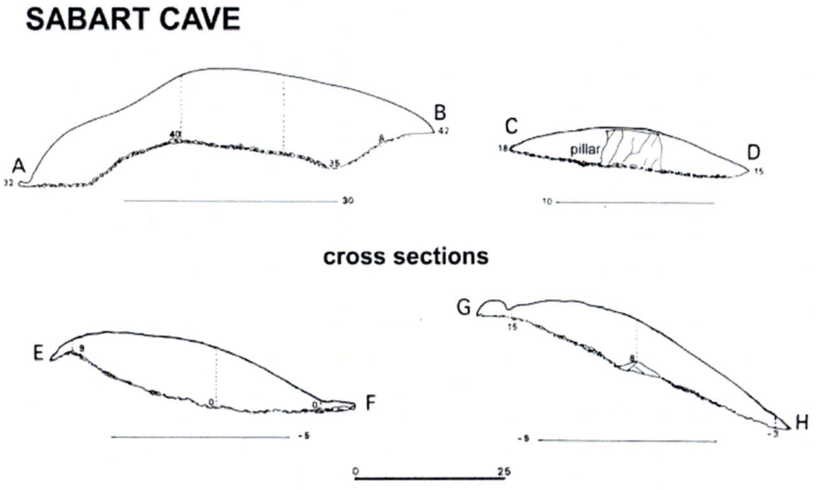

Fig. 2.177 Cross sections of Big Chamber

- at the west, a room with a domed ceiling occupied at its center by a cone of scree.
- at the east, a gallery, wide but of low height, inclined toward the east.

A big block separates these two parts. This block is actually an ancient rock pillar currently disengaged from the ceiling. To the west of the hall, a wide, sloping gallery, the Great Gallery, provides access to the Renouveau Chamber and to galleries that communicate with the outside. In the eastern part, a narrow passage between unstable blocks joins the Petit-Pousail Cave. The floor of the Big Chamber is completely masked by a scree whose blocks are more or less covered with flowstone. There are some beautiful stalagmites south of the large block. The ceiling has a rounded shape (Fig. 2.177).

The Renouveau Hall (Fig. 2.172) is very different from the previous one. It is 210 m long and 90 m wide. It is actually also a large gallery. The floor is a sandy zone at its lowest part, at the entrance and a scree for the rest of the chamber. To the south, the gallery ends on a blockfield of enormous rocks which forms a very steep slope at the top of

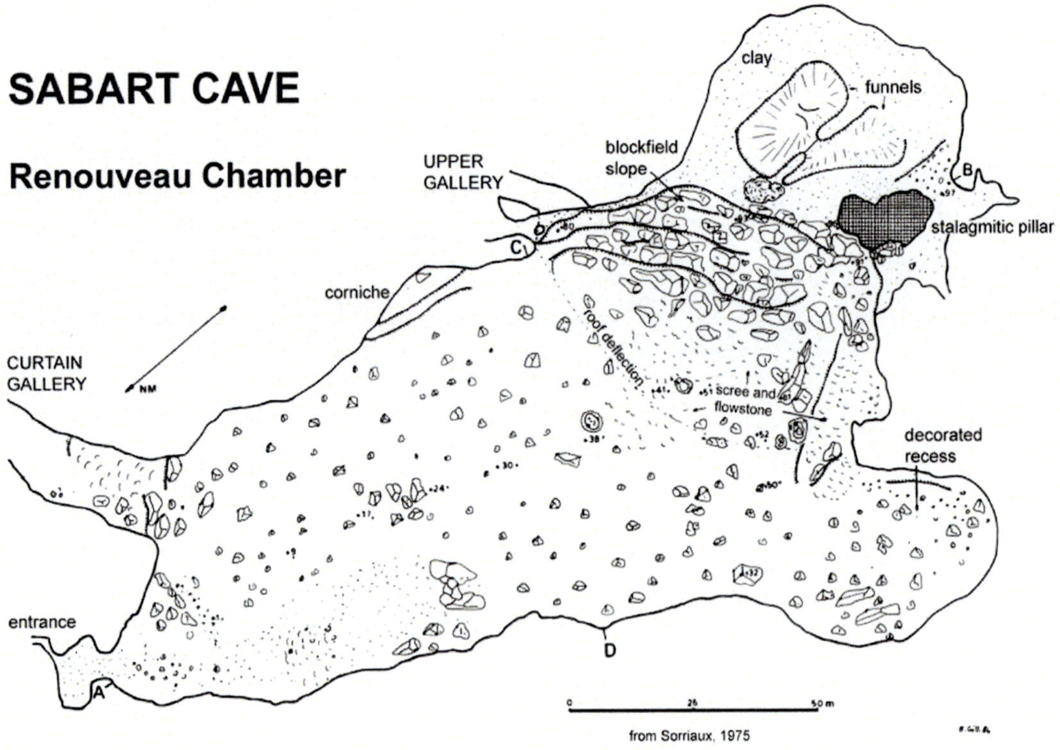

Fig. 2.178 Map of Renouveau Chamber

Fig. 2.179 Longitudinal section of Renouveau Chamber

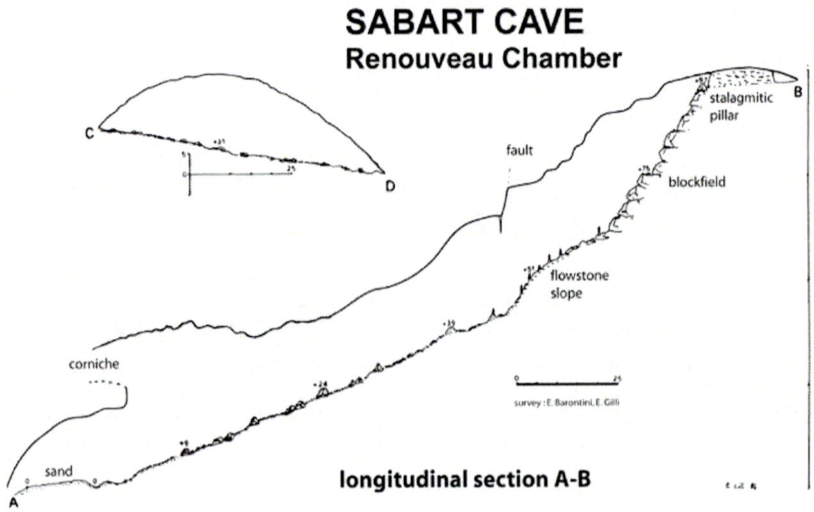

which is the highest point of the room, which is 93 m above the entrance level. This high point is a flat area whose center is occupied by two clay funnels. From this flat, to the northeast there is a large Upper Gallery and to the southwest a short cave network with speleothems.

To the west, at the foot of the big blockfield, the gallery forms a large niche with speleothems. East of the entrance, a gallery decorated with rimstones goes down to the northeast; it extends parallel to the Great Gallery and the Upper Gallery (Fig. 2.174). The ceiling of the room is an arch in the lower part of the cavity; it is about 20 m high. Beyond, the ceiling rises suddenly to reach the flat area.

Both rooms are totally devoid of water circulation.

2.24.2 Geology and Hydrogeology

The Sabart Cave is located in the northern end of the rocky outcrop of Cap-de-Lesse, at the confluence of Vicdessos and Ariège Valleys, 560 m.a.s.l. This spur is traversed from one end to the other by a series of karst galleries that form the Niaux-Lombrives-Sabart Cave network. The sediment filling of these caves was studied in a thesis work to define the direction of the water flows at different times (Sorriaux 1982).

The Big Chamber extends parallel to the slope. Its north wall is about 75 m from the outside. Above the ceiling, the thickness of the limestone is about 20 m. The Renouveau Chamber Room is located under a cliff, the Saut-de-l'Ours (Bear's Jump), which is a fault escarpment. The thickness of the limestone above the ceiling is about 75 m at the highest point of the hall and 120 m at its lowest point.

The Sabart Cave is dug in the thick series of Barremo-Aptian massive limestone on the northern flank of the Sabart-Cap-de-Lesse anticline. This large fold is located at the southern end of the sedimentary basin of Tarascon-sur-Ariège, north of the Hercynian massif of Miglos. The Sabart Formation in which both chambers are dug is a micritic compact gray limestone with some dolomitic levels. The thickness of the layers varies from a few decimeters to a few meters. The entire limestone series has a thickness of 300 m.

In Big Chamber, the layers are almost vertical in the western part of the room. The strike is N180°. The dip decreases to the east and reaches 15° east at the eastern end of the hall (Fig. 2.176). The fractures are numerous. They form a regular grid over the entire cave which locally makes possible the confusion with the bedding planes, especially as sometimes the interbeds relay the joints. The direction of the main joint set is N110°. Many faults cut the room, mainly in its northern part, where an E-W fault inclined 70° toward the north is clearly visible (Fig. 2.175).

For Renouveau Chamber, only a summary geological survey was made. The northern part of the hall seems to be hollowed out in series inclined 50° to the north. The series then plunges to the south and thus forms an anticline whose axis corresponds substantially to the place where the ceiling inclination increases. In this area, an important fracture seems to influence the morphology of the room (Fig. 2.179). In the flat area, at the top of the blockfield slope, the layers have an N20° to N30° strike with a dip of 30° to the east.

Cave filling is varied. In the western part of Big Chamber, the floor is masked by a blockfield whose elements are of metric to decimetric size. The southern border of the blockfield is covered with flowstone. In the eastern part of the hall, the blockfield gives way to a finer scree, covered with speleothems in the lower part of the room. Along the southern edge, the filling is clay. At the northeast corner of the hall, a fine and earthy scree cone is fed by the material coming from the Petit-Pousial Cave. At certain points of the cave, the presence of crystalline pebbles stuck on to the roof indicates that it was totally clogged at the glacial times and then cleared later by strong circulations of water.

The artificial entrance of Renouveau Chamber is dug into the sandy mass that forms the floor of the Big Chamber in its northwestern part. Most of the floor is formed by a blockfield locally covered with flowstone. The blocks are of variable size. The scree located in the southern bottom of the room is a blockfield of huge elements stuck together by speleothems. The upper flat area is very clayey.

2.24.3 Genesis and Stability

The Sabart Cave is a giant karst drain that is today no more functional. The current shape is the result of the successive collapses of the ceiling, probably linked to the readjustment of the constraints in the massif during the deepening of the Ariège and Vicdessos Valleys. The presence and the size of crystalline pebbles brought by the waters and still stuck in certain points to the roof testify to the importance of the water circulations which crossed this gallery. The central block is an old rocky section that separated two parallel galleries and is now detached from the ceiling.

Both chambers are stable. In Big Chamber, only the wall located north of the pillar shows instability signs. The rock is disorganized there. It is a relaxation phenomenon comparable to that which degraded the central pillar of the ceiling. It was probably caused by the proximity of the outside slope. The rest of the cavity seems stabilized. The importance of the flowstone deposit in the room leads one to think that its stability has been acquired for a long time. However, it must be taken into account that the growth speed of speleothems may be very fast.

2.25 Tindoul de la Vayssière Shaft (Salles-la-Source, Aveyron, France)

2.25.1 Description

See Figs. 2.180, 2.181, 2.182.

Known since always, the Tindoul-de-la-Vayssière Shaft was one of the first chasms explored in France. It was indeed visited in 1785 by Father Carnus a professor at the Royal College of Rodez. The real explorations did not begin until 1890. In 1891, the French caver Gaupillat installed a staircase whose remains are still visible (Minvielle 1977). The cave system extends on more than 2 km (Bigot 2004).

The shaft is located north of the city of Rodez. It is reached by the D988 to Sebazac-Concourès, then by the D904 for 6 km. There, a sign indicates on the left the direction of the cave. A 400-m-long track leads to a bunch of trees behind which is the shaft (coordinates: 44° 26′ 09″ N, 2° 34′ 33″ E).

This is a big collapse shaft. The oval entrance measures 40 m by 25 m. A 40-m-deep pitch leads to the top of a scree slope. A recess 30 m long and 20 m wide abuts to the east on a wall at the foot of which a narrow passage between the blocks leads to a dry gallery and then to an underground river. The north and south walls of the shaft are very fractured.

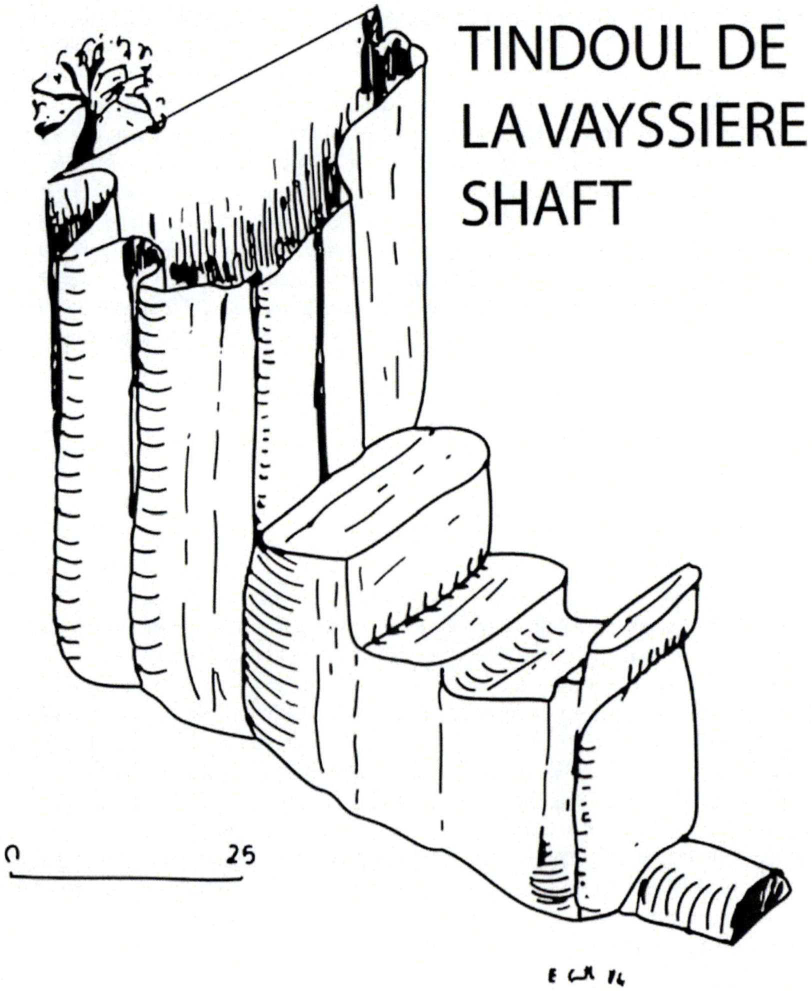

Fig. 2.180 3D sketch of Tindoul-de-la-Vayssière Shaft

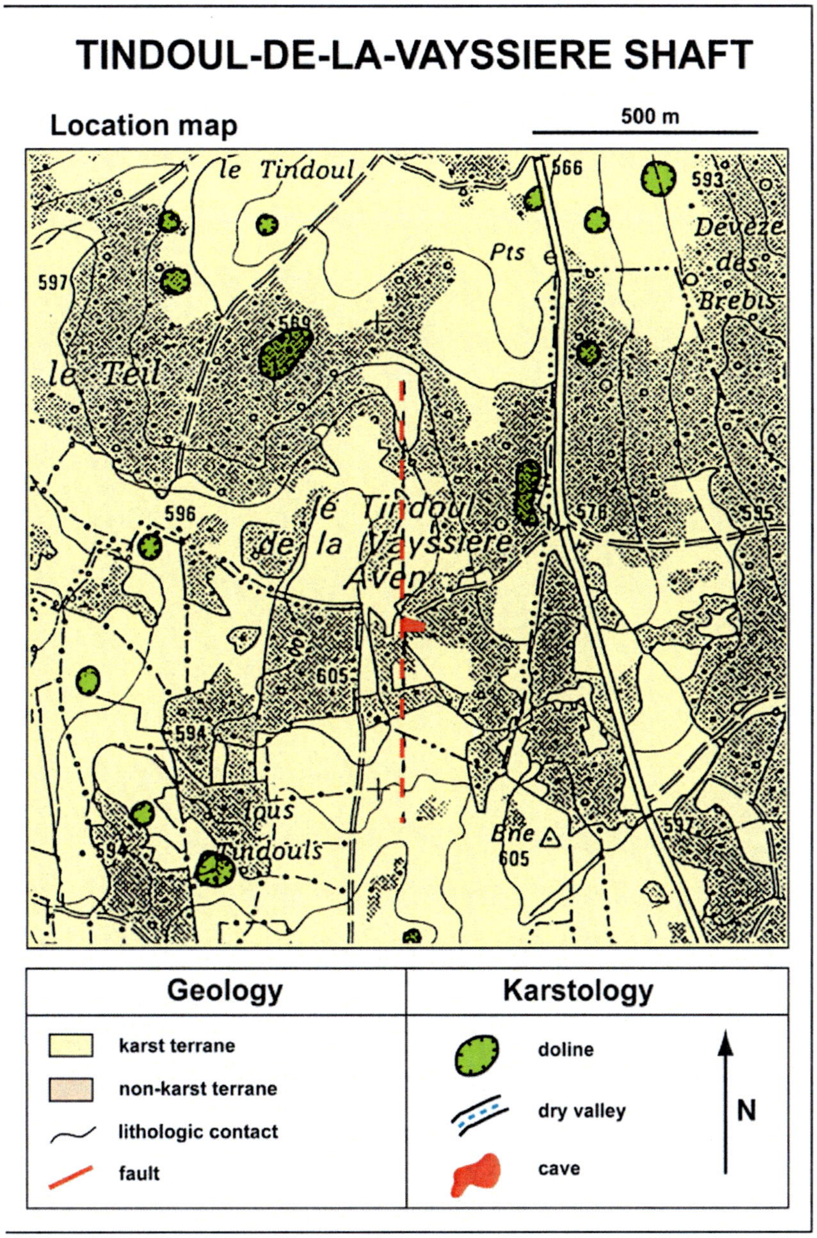

Fig. 2.181 Geomorphological map of Tindoul-de-la-Vayssière Shaft

2.25.2 Geology and Hydrogeology

The chasm is at an altitude of 575 m in the central part of the Causse-du-Comtal karst plateau. The place is covered with a sparse forest, and its surface is dotted with sinkholes. The cultivable soil is reduced and partially covers fields of lapies. The thickness of the limestone above the shaft recess is about 30 m. The chasm is located in a sub-horizontal limestone series, framed north and south by two vertical E-W faults.

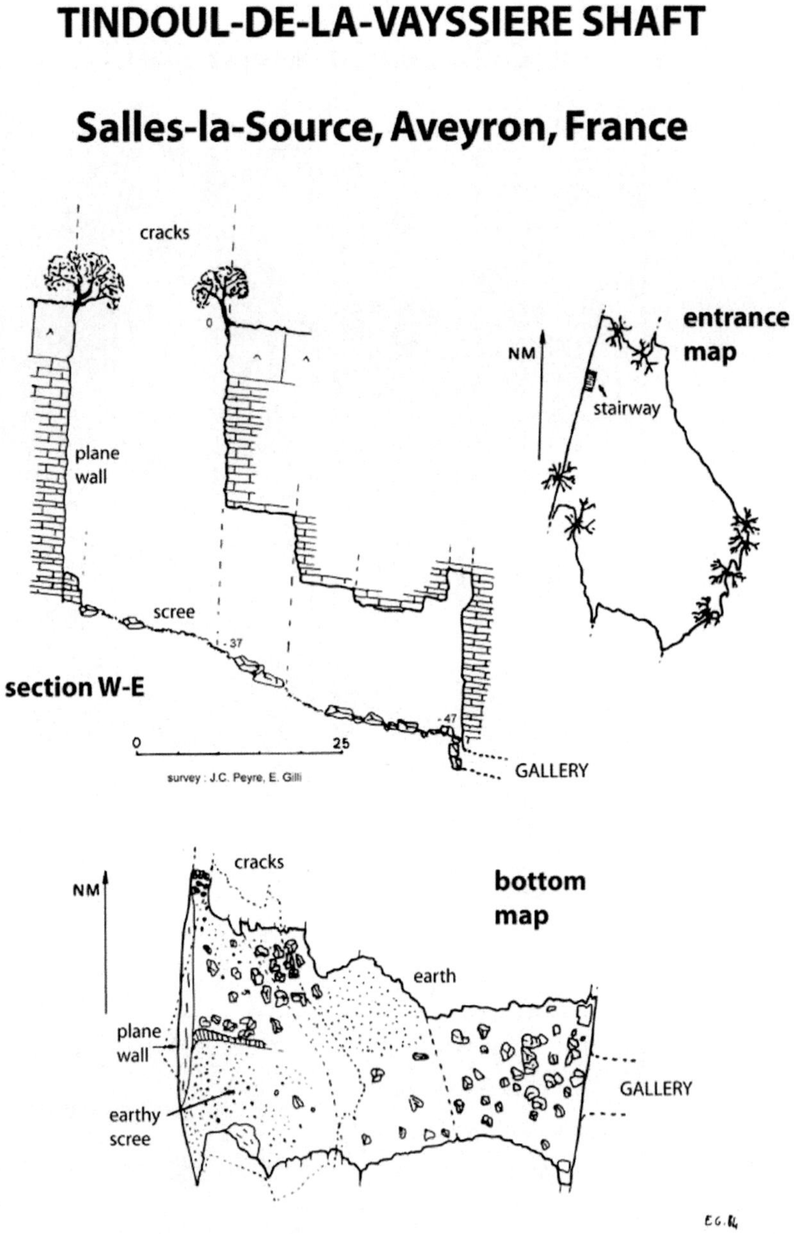

Fig. 2.182 Map and cross section of Tindoul-de-la-Vayssière Shaft

From the entrance to the bottom, the series is as follows:

- 5 m of pink crystalline dolomite, forming a large bench;
- 50 m of compact limestone in 0.4- to 0.8-m-thick beds separated by fine interlayers of laminated marly limestone.

Both formations are aged Bajocian and Aalenian. The hydrologic base level is formed by Toarcian marls and black schistose clays. The layers have an N-S strike with a dip of 6° to the east.

In the cave, a dense network of N10-15° vertical cracks is observed. The eastern and western walls of the chasm are formed by the walls of two of these fractures. These are probably faults although the observation does not allow being affirmative (no striation, no crushed zone).

The floor of the entrance pitch is covered with an earthy scree with blocks of all sizes. This scree was formed by the collapse of the roof during the genesis of the entrance pitch. It is also fed by the small blocks detached from the walls under the effect of external agents (cryoclasty, decompression).

The entrance pitch is located on the path of an underground river that can be followed toward upstream for about 700 m. This river sinks into a scree, 300 m before arriving at the shaft. The water emerges at Salles-la-Source 3 km west of the cave.

2.25.3 Genesis and Stability

It is a typical collapse shaft formed by a roof collapsing in a gallery crossed by an underground river in a very fractured zone. The collapse created a dam. The river is now using new karst conduits to bypass this dam. The north and south walls are highly fractured and very unstable, especially in the western half of the chasm. Currently, the scree does not evolve. The river no longer bathes the scree as it disappears in the upstream gallery 300 m before reaching the shaft zone.

2.26 List of Participants to the Survey Works

- Armand Shaft: E. Gilli and J.C. Peyre.
- Baume Shaft: D. Bessaguet, E. Gilli and J.C. Peyre from the 1970 map by Chapuis et al. (1971).
- Bédeilhac Cave: C. Adam de Villiers and E. Gilli from the 1982 map by Spéléo Club du Haut Sabarthez.
- Bournillon Cave: J.P. Bozonet, E. Gilli and P. Nigon from GSV (1975).
- Castelette Cave: C. Adam de Villiers, A. Bocuse, B. Bocuse, M. Mellouli and E. Gilli, from SCM (1982).
- Cennet ve Cehennem Shafts: E. Gilli and J.P. Mairetet from Aygen (1970) and Gilli (1982).
- Champclos Fountain: E. Gilli and J.P. Mairetet from Chabaud and Rivol (1976).
- Coufin Cave: E. Gilli and P. Nigon from GSV (1980).
- Fouillac Shaft: E. Gilli, B. Thérond and R. Thérond.
- Kocain Cave: N. Boullier, C. Chabert, F. Duverneuil, E. Gilli and J.P. Mairetet from Chabert (1979).
- Mas d'Azil Cave: C. Adam de Villiers, J.F. Cottin, J. P. Mairetet and E. Gilli from a map by Direction départementale de l'Equipement.
- Moras Cave: J.P. Bozonet, R. Thérond and E. Gilli.
- Mort Ru Cave: J.P. Bozonet and E. Gilli from SCS (1977) and Talour (1975).
- Aven Noir: E. Gilli, from Caubel's map (SCC 1978).
- Aven d'Orgnac Shaft: E. Gilli and J.C. Peyre from the map by Groupe Spéléo de Vallon-Pont-d'Arc (Trébuchon 1967).
- Padirac Shaft: E. Gilli and J.P. Mairetet.
- La Verna Chamber in Pierre Saint-Martin Shaft: T. Gaschat, E. Gilli and R. Thérond from two maps by Brook (1974) and Association des Recherches Spéléologiques Internationales de la Pierre Saint-Martin (ARSIP 1984).
- Barrenc du Pla de Périllos Shaft: C. Adam de Villiers and E. Gilli from Salvayre (1977).
- Poudrey Shaft: E. Gilli, from the map by the Gouffre de Poudrey Office.
- Pradières Cave: map by the Spéléo Club du Haut Sa'barthez (Claustres 1982).
- Ravières Shaft: D. Bessaguet, E. Gilli and C. Peyre.
- Réveillon Shaft: E. Gilli and J.P. Mairetet.
- Riusec Cave: C. Chabert, E. Gilli, J.P. Mairetet, M. Mellouli and R. Thérond.
- Sabart Cave: C. Chabert, J.F. Cottin, E. Barontini and E. Gilli from the map by Spéléo Club du Haut Sabarthez (Claustres 1982).
- Tindoul-de-la-Vayssière Shaft: E. Gilli and J.C. Peyre.
- Sarawak Chamber (Borneo): F. Verlaque, R. Schejbal, Th. Gaschat, P. Delange, E. Gilli.

References

André, D.: Aven armand. Spéléo Mag. **97–98**, 86–91 (2017)

ARSIP: Le karst de la Pierre Saint Martin. Assoc des recherches spél. de la Pierre Saint Martin, Arette (1984)

Aygen, T.: Les Gouffres du Paradis et de l'Enfer et la grotte de Narlikuyu. Tourist Booklet, Silifke Turizm Denergi (1970)

Azéma, M., Brasier, L.: Le beau livre de la préhistoire: De Toumaï à Lascaux 4, Dunod, p. 420 (2016)

Bigot, J.-Y.: Spéléométrie de la France, Spelunca mémoires, vol. 27, p. 160 (2004)

BRGM.: Atlas des potentialités aquifères des formations pyrénéennes projet POTAPYR—BRGM/RP-66912-fr (2017)

Brook, D.: A survey of La Verna, British Cave Research Association Bull, vol. 4 (1974)
Caillault, S., Dominique Haffner, D., Krattinger, T., Delannoy, J.-J.: Spéléo Sportive dans le Vercors-Tome 2, p. 160. Edisud, Aix-en-Provence (1997)
Caillault, S., Dominique Haffner, D., Krattinger, Th, Delannoy, J.-J.: Spéléo Sportive dans le Vercors-Tome 2, p. 208. Edisud, Aix-en-Provence (1999)
Cavers, N., André D., Marbach, G.: Les splendeurs cachées de l'aven Noir. Spéléo, **58**, 40 (2007)
Chabaud, M., Rivol, D.: La fontaine de champclos. Spelunca **4**, 173 (1976)
Chabert, C.: La grotte de Kocain, Grottes et Gouffres, Bull. S.C. Paris 72 (1979)
Chapuis, C., Jacquot, Picquard, Petrequin, P.: Le gouffre de la Baume. Bull. Assoc. Spéleo. de l'Est **8**, 34–36 (1971)
Claustres, M.: La grotte de Pradières, Caougno. Bull. Spéléo Club du Haut Sabarthez, Tarascon-sur-Ariège **12**, 49 (1982)
Delannoy, J.-J., Gauchon, C., Jaillet, S.: L'Aven d'Orgnac, valorisation touristique, apports scientifiques, coll. « Edytem » **5**, 185 (2007)
Delannoy, J.-J., Gauchon, C., Hobléa, F., Jaillet, S., Maire, R., Perrette, Y., Perroux, A.-S., Ployon, E., Vanara N.: Le karst: des archives paléogéographiques aux indicateurs de l'environnement, Géomorphologie: relief, processus, environnement, Paris, **15–2**, 84–85 (2009)
Dubrana, D.: Le gouffre de Padirac. Découvertes Gallimard, p. 128 (2013)
Gèze, B.: Carte géologique de la France, St Affrique 1/80.000°, B.R.G. M., Orléans (1965)
Gèze, B.: Le principe de l'inversion de relief en région karstique, 5th G. I.S. Stuttgard M, 20/1 (1969)
Gèze, B., Cavailié, A.: Guides géologiques régionaux. Aquitaine Orientale, Masson edit (1977)
Gigon, R., Monnin, J.: Inventaire Spéléologique du S.E. du département du Doubs, Ann. Spéléo. XXI-1 (1966)
Gilli, E.: Les gouffres du Paradis et de l'Enfer, Spéléologie, Bull. Club Martel C.A.F. Nice, 119 (1982)
GSV: Plan du Bournillon Bull. Groupe Spéléo Valentinois (1975)
GSV: Plan de la grotte de Coufin. Bul. Groupe Spéléo Valentinois (1980)
Harper, K.: The fate of Rome. Climate, Disease and the End of an Empire. Princeton University Press, p. 418 (2017)
Le Bec, M.: Un barrage au centre de la Terre. H2O Magazine. http://www.h2o.net/infrastructures-developpement-durable/un-barrage-au-centre-de-la-terre.htm (2008). Last accessed 28 Jan 2020
Loiseleur, B.: Le massif du Seuil (Chartreuse, France): organisation des réseaux souterrains. Karstologia **24**, 13–28 (1994)
Martel, E.A.: Les abîmes, p. 580. Delagrave edit, Paris (1894)
Martel, E.A.: Aven Armand, p. 48. Artières et Maury edit, Millau (1937)
Minvielle, P.: Séisme et Spéléologie. Spelunca **1967**(4), 295–296 (1967)
Minvielle, P.: Grottes et Canyons, Denseil edit., p. 234 (1977)
Moret, L.: Rapport sur les conditions géologiques de la chute de la grotte du Bournillon, projetée dans la vallée de la Bourne (Vercors). Rapport E.D.F, Grenoble (1950)
Pétris, J.-J.: L'occupation des grottes de Bédeilhac et du Mas d'Azil durant la Seconde Guerre mondiale, pp. 201–233. Archives ariégeoises, Saint-Girons (2017)
Prost, S., Reilé, P., Villegas, J.-P.: Le réseau du Verneau (Doubs). Spelunca **148**, 21–29 (2017)
Rauzy, C.: La grotte de Bédeilhac. Caougno, Bull du S.C. du Haut Sabarthez, Tarascon-sur-Ariège **12**, 30–31 (1982)
Renault, P.: Contribution à l'étude des actions mécaniques et sédimentologiques dans la spéléogénèse. Ann. Spéléo. CNRS, Paris 23–1 (1968)
Renault, P.: Les rapports entre dolines et galeries de grottes d'après quelques exemples du Causse de Gramat (Lot). Bull. Assoc. Géogr. Fr. 389–390 (1971)
Renault, P.: Études récentes sur le karst de Niaux-Lombrive-Sabart (Ariège). Karstologia **2**, 17–22 (1983)
Rouire, J., Rousset, C.: Guide géologique régionaux. Causses Cevennes Aubrac, Masson edit (1973)
Salvayre, H.: Contribution à l'étude des origines des résurgences côtières de Font Estramar, Font Dame (Massif des Corbières—Pyrénées Orientales—France). Mémoires et Documents, CNRS edit., Phénomènes karstiques, 1974–15-II (1974)
Salvayre, H.: Spéléologie et hydrogéologie des massifs calcaires des Pyrénées Orientales, p. 250. Spéléo. Conflent, special issue, Inv (1977)
Sorriaux, P.: Les remplissages du système Niaux-Lombrive-Sabart, Thèse 3ème cycle géol., Toulouse (1982)
SCC: Spéléo Causse Noir 2. Spéléo Club des Causses, Millau (1978)
SCM: Plan de la grotte de la Castelette. Bull. speleo club de Marseille, CAF (1982)
SCS.: La grotte du Mort Ru. In: Grottes de Savoie, Bull. speleo club de Savoie (1977)
Talour, B.: Inventaire spéléo du Massif de la Chartreuse (1975)
Trébuchon, J.C.: Le nouveau réseau de l'Aven d'Orgnac. Spelunca **1967**(1), 7–19 (1967)
Trombe, F.: Grotte supérieure de Ruisec, Ann. Spél. Fr., CNRS. II-2-3 (1947)
Trombe, F., Dresco, E., Halbronn, G., Henry-La-Blanchetais, Ch., Negre, J.: Recherches souterraines dans les Pyrénées centrales, années 1945 à 1947. Ann. Spélé. Fr., CNRS edit. **II**, 2–3 (1947)
Varrel, E.: Le clone numérique de la salle de la Verna, Spelunca. **98**, 25–42 (2005)

3 Comparative Analysis and Synthesis

This chapter is a synthesis of the observations made in the different caves studied in Chap. 2. Large chambers are close to the surface. Most of the examples are stable despite their size that may reach several tens of meters in width. Cave ceilings tend to reach an equilibrium shape which is most often a rounded vault, but several examples have a flat shape. The thickness of the limestone beds is an important parameter of stability. Fracturing also plays an important role. A high density of fractures is a cause of unstability, but conversely, large fractures may form stable surfaces. Most caves are located in areas where rainfall is important which emphasizes the role of water. Three modes of natural digging are considered: dissolution, dissolution and collapse, scouring-racking and collapse. The chapter ends with the conclusion of the thesis work where it is suggested that the large underground volumes correspond to the removal, by the water, of the decompressed rocks located beneath a pre-existing natural vault. The large volumes could, therefore, be prefigured inside the massif.

3.1 Overview—Summary Table

All studied caves are grouped in tables that present their different parameters (Tables 3.1, 3.2, 3.3, 3.4, 3.5).

Symbol:

Each cavity is represented by a symbol which will be used systematically in the text of the third and fourth parts.

Nature:

C = chamber; G = giant gallery; S = shaft; P = porch.

Dimensions (Table 3.1):

- width
- length (the number indicates the maximum dimensions)
- height

Ceiling (Table 3.1):

There are three categories:

- rounded (Verna)
- corbelled (Armand)
- flat (Poudrey).

Topography (Table 3.1):

- Inclination of the roof:

The vault is sometimes flat and inclined; the number indicates the average value of this inclination in degrees.

- Thickness of the roof:

This is the minimum thickness of land above the cavity.

- Distance to slope:

The number indicates the distance between the cave and the nearest slope.

Filling (Table 3.2):

The table indicates the nature of the floor of the cavity.

Speleothems (Table 3.2):

This concerns the speleothems located on the ceiling and walls (stalactites and flowstones).

Regional Geology (Table 3.3)

E. Gilli, *Big Karst Chambers*, Advances in Karst Science,
https://doi.org/10.1007/978-3-030-58732-1_3

Table 3.1 Dimensions and topographic context

Name	Text symbol	Nature	Dimensions			Ceiling				Topography	
			Width	Length	Height	Plane	Corbelled	Vault	Ceiling inclination	Thickness above ceiling	Distance from slope
Armand	Ar	C	60	85	40		●		20°	45	–
Baume	Ba	C	70	120	70		●		–	50	–
Bédeilhac-Rolland Grave	Be	G	65	65	8	●		●	–	280	–
Bournillon	Bo	P	60	165	105		●		5°	350	0
Castelette	Ca	C	55	110	8	●			20°	30	150
Cehennem	Cm	S	50	90	100			●	–	0	–
Cennet	Ct	G	50	250–500	50?			●	–	50	
Champclos	Ch	C	45	50	35			●	–	55	80
Coufin	Co	C	45	55	15	●			0°	260	3
Fouillac	Fo	C	80	110	18			●	–	15	–
Kocain	Ko	G-C	150	450	80			●	–	?	–
Mas d'Azil	Ma	G	100	450	46	●	●		15°	60	–
Moras	Ms	C	60	75	20		●		–	30	–
Mort Ru	Mr	C	50	90	32			●	–	25	30
Noir	No	C.G.S	65	200	28	●			0°	15	–
Orgnac De Joly Hall	Or	C	80	115	30	●	●		10°	15	–
Padirac	Pa	S	50	50	45				–	0	–
Pierre St. Martin La Verna	Pi	C	230	270	180			●	–	100	–
Pla de Perillos	Pe	C	75	75	24			●	–	20	
Poudrey	Po	C	100	130	37	●			5°	25	–
Pradières	Pr	G	75	200	17			●	–	20	<100
Ravières	Ra	C	65	65	13		●		–	0	–
Réveillon	Re	P	35	85	40	●	●		–	30	0
Riusec A Hall	Ri	C	95	120	30	●			30°	100	80
Riusec Trombe Hall	Rt	C	90	125	40	●			45°	80	20
Sabart Great Chamber	Sg	G	100	200	22			●	–	20	75
Sabart Renouveau	Sr	C	100	205	30			●	–	75	–
Tindoul	Ti	S	30	0	20–38					30	

C = chamber; G = gallery; P = porch; S = shaft

Three categories were distinguished:

- very tectonized regions (many folds, thrust, large fault, e.g., Pyrenées);
- folded regions (e.g., Jura);
- plateaus (e.g., Grands Causses).

Lithology (Table 3.3)

- Nature of the rock:

The table shows the series encountered from the top of the cave to its base.

Table 3.2 Cave fillings

Name	Text symbol	Nature	Filling			Speleothems		
			Clay	Rocks	Flowstone	Rare	On cracks only	Everywhere
Armand	Ar	C						
Baume	Ba	C	●	●		●		
Bédeilhac-Rolland Grave	Be	G	●			●		
Bournillon	Bo	P		●		●		
Castelette	Ca	C	●	●		●	●	
Cehennem	Cm	S		●				●
Cennet	Ct	G	●	●		●		
Champclos	Ch	C			●			●
Coufin	Co	C	●					●
Fouillac	Fo	C		●	●			●
Kocain	Ko	G-C	●	●	●			●
Mas d'Azil	Ma	G		●			●	
Moras	Ms	C		●			●	
Mort Ru	Mr	C		●		●		
Noir	No	C.G.S		●		●		
Orgnac De Joly Hall	Or	C		●	●			●
Padirac	Pa	S		●		●		
Pierre St. Martin La Verna	Pi	C		●		●		
Pla de Perillos	Pe	C	●					●
Poudrey	Po	C	●	●	●			●
Pradières	Pr	G		●	●			●
Ravières	Ra	C		●		●		
Réveillon	Re	P		●		●		
Riusec A Hall	Ri	C		●	●	●		
Riusec Trombe Hall	Rt	C		●		●		
Sabart Great Chamber	Sg	G		●	●			●
Sabart Renouveau	Sr	C	●	●	●			●
Tindoul	Ti	S						

C = chamber; G = gallery; P = porch; S = shaft

- Age of the rock:

  Unit is a million years old.

- Beds thickness:

  Three classes were distinguished:

- very thick beds more than 0.5 m,
- thick benches between 0.2 and 0.5 and,
- thin beds less than 0.2 m.

- Nature of the joints:

  This section illustrates the cohesion of limestone beds:

- weak cohesion beds, (e.g., with a marly joint);
- stuck beds (massive limestone).

- Dip:

  This is the dipping of the layers, classified into three categories:

- reduced dip less than 10°;
- medium dip less than 45°;
- important dip higher than 45°.

Fracturing (Table 3.3):

- Major fracture.

Table 3.3 Geological context

Name	Text symbol	Nature	Regional geology			Lithology										Fracturation			
						Nature of rock	Rock age	Beds thickness			Limestone bed joints		Dip			Presence of a main fracture	Number of sets		
			Tabular	Folded	Intense tectonics			<0.2	0.2–0.5	>0.5	Cemented	Weak	<10°	10°–45°	>45°		0	1	N
Armand	Ar	C	●			Sublithographic limestone	●	●			●		●		●	●	●		
Baume	Ba	C		●		Compact white limestone, sandy limestone, marls	150			●	●		●			●		●	
Bédeilhac-Rolland Grave	Be	G			●	Urgonian massive limestone	110			●	●			●				●	●
Bournillon	Bo	P	●			Sandy limestone, Urgonian massive limestone	115	●		●	●			●		●			
Castelette	Ca	C			●	Dolomitic limestone	145			●		●		●			●		
Cehennem	Cm	S	●			Microfractured or brecciated limestone	?			●	●		●			●		●	
Cennet	Ct	G	●			Microfractured or brecciated limestone	?			●	●		●			●		●	
Champclos	Ch	C	●			Marly limestone and marls	160	●					●				●		
Coufin	Co	C	●			Lower Urgonian red sandy limestone	115		●		●		●					●	
Fouillac	Fo	C	●			Compact reef limestone	150			●	●					●			●
Kocain	Ko	G-C		●		Microfractured massive limestone	?			●	●			●		●			●
Mas d'Azil	Ma	G		●		Massive limestone, clayey knobbly limestone	55			●			●						●
Moras	Ms	C	●			Sandy limestone,	75			●		●	●			●		●	
Mort Ru	Mr	C			●	Lower Urgonian red sandy limestone	115	●				●	●			●			●
Noir	No	C.G.S		●		Small bed dolomite and limestone, sublithographic platy limestone	170	●	●	●		●	●			●			●
Orgnac De Joly Hall	Or	C	●			Urgonian limestone	115			●	●		●			●			●
Padirac	Pa	S	●			Lithographic limestone, few marls layers, sub-oolitic limestone	170			●	●		●			●			●

(continued)

Table 3.3 (continued)

Name	Text symbol	Nature	Regional geology			Lithology										Fracturation			
						Nature of rock	Rock age	Beds thickness			Limestone bed joints		Dip			Presence of a main fracture	Number of sets		
			Tabular	Folded	Intense tectonics			<0.2	0.2–0.5	>0.5	Cemented	Weak	<10°	10°–45°	>45°		0	1	N
Pierre St Martin La Verna	Pi	C	●		●	Very massive gray limestone, shale, sandstone, black limestone with white veins	100	●		●	●	●	●		●	●			●
Pla de Perillos	Pe	C			●	Massive and compact sublithographic limestone	115		●		●			●		●			●
Poudrey	Po	C	●			Massive limestone, marls, marly limestone	150		●	●	●		●			●			●
Pradières	Pr	G			●	Urgonian semi-crystalline limestone	110			●	●			●		●			●
Ravières	Ra	C		●		Sublithographic limestone in small beds then massive beds, marls layers	150		●	●		●	●			●			●
Réveillon	Re	P	•			Sublithographic limestone, massive sandy limestone, sandy limestone in small beds, marls	175		●	●		●		●		●			●
Riusec A Hall	Ri	C			●	Urgonian limestone, small beds of limestone and marls, massive black limestone	100–120	●		●	●	●			●	●			●
Riusec Trombe Hall	Rt	C			●	Idem	100	●		●	●	●			●	●			●
Sabart Great Chamber	Sg	G			●	Compact fine-grained limestone	110			●	●				●	●			●
Sabart Renouveau	Sr	C			●	Idem	110			●	●				●	●			●
Tindoul	Ti	S	●			Massive dolomite, compact limestone	170					●						●	●

C = chamber; G = gallery; P = porch; S = shaft

Table 3.4 Hydrology and climate

Name	Text symbol	Nature	Current hydrology			Current climate	
			Dry	Wet	River	Glacial	Tempered
Armand	Ar	C		●		●	
Baume	Ba	C			●		●
Bédeilhac-Rolland Grave	Be	G	●				●
Bournillon	Bo	P			●	●	
Castelette	Ca	C			●		●
Cehennem	Cm	S	●				●
Cennet	Ct	G			●		●
Champclos	Ch	C			●		●
Coufin	Co	C			●	●	
Fouillac	Fo	C	●				●
Kocain	Ko	G-C	●				●
Mas d'Azil	Ma	G			●		●
Moras	Ms	C		●		●	
Mort Ru	Mr	C	●			●	
Noir	No	C.G.S		●			●
Orgnac De Joly Hall	Or	C	●				●
Padirac	Pa	S			●		●
Pierre St. Martin La Verna	Pi	C			●	●	
Pla de Perillos	Pe	C	●				●
Poudrey	Po	C		●			●
Pradières	Pr	G	●			●	
Ravières	Ra	C		●			●
Réveillon	Re	P			●		●
Riusec A Hall	Ri	C		●			●
Riusec Trombe Hall	Rt	C		●			●
Sabart Great Chamber	Sg	G	●	●			●
Sabart Renouveau	Sr	C	●	●			●
Tindoul	Ti	S			●		

C = chamber; G = gallery; P = porch; S = shaft

This column indicates the presence of a major visible fracture (fault, diaclase).

- Number of cracks sets (fractures):

Three categories were distinguished:

- caves without fractures,
- caves with one fracture network and,
- caves with several fracture networks.

Current hydrology (Table 3.4):

This topic states

- the presence of an underground river (river),
- the presence of diffuse (wet) runoff and,
- total absence of runoff (dry).

Current climate (Table 3.5):

The cavities undergoing the action of the gel were opposed to the others.

Genesis (Table 3.5):

This section groups together the different types of digging defined in Chapter 7 of the third part.

Table 3.5 Genesis and stability

Name	Text symbol	Genesis					Unstability		
		Dissolution	Collapse	Mixt	Suffosion	Scouring	Ceiling	Walls	Floor
Armand	Ar				●				
Baume	Ba			●					
Bédeilhac-Rolland Grave	Be	●							
Bournillon	Bo			●				●	
Castelette	Ca		●				●		
Cehennem	Cm		●					●	
Cennet	Ct			●					
Champclos	Ch		●			●		●	
Coufin	Co			●					
Fouillac	Fo		●					●	
Kocain	Ko			●					
Mas d'Azil	Ma					●		●	
Moras	Ms		●					●	
Mort Ru	Mr		●				●	●	
Noir	No			●		●			
Orgnac De Joly Hall	Or			●				●	
Padirac	Pa		●						
Pierre St. Martin La Verna	Pi			●	●			●	●
Pla de Perillos	Pe			●					
Poudrey	Po		●		●			●	●
Pradières	Pr			●					
Ravières	Ra					●		●	●
Réveillon	Re					●		●	
Riusec A Hall	Ri				●			●	●
Riusec Trombe Hall	Rt				●			●	●
Sabart Great Chamber	Sg			●					
Sabart Renouveau	Sr			●					
Tindoul	Ti		●						

C = chamber; G = gallery; P = porch; S = shaft

Unstability (Table 3.5):

The table shows

- the presence of withdrawal signs on the floor,
- the presence of ready to collapse blocks on the walls and,
- the presence of ready to collapse blocks on the roof, or a ceiling already collapsed.

3.2 Classification of Caves

The following symbols are used for the different caves: Armand (Ar), Baume (Ba), Bedeilhac (Be), Bournillon (Bo), Castelette (Ca), Cehennem (Ce), Cennet (Ct), Chamclos (Ch), Coufin (Co), Fouillac (Fo), Kocain (Ko), Mas d'Azil (Ma), Moras (Ms), Mort Ru (Mr), Noir (No), Orgnac (Or), Padirac

(Pa), Pierre St-Martin Verna chamber (Pi), Pla de Perillos (Pe), Poudrey (Po), Pradières (Pr), Ravières (Ra), Réveillon (Re), Riusec A hall (Ri), Riusec Trombe hall (Rt), Sabart great hall (Sg), Sabart Renouveau chamber (Sr) and Tindoul (Ti).

The caves are classified in the four categories defined in the first part of Chap. 2: the giant galleries, the chambers, the collapse shafts and the porches.

- Giant galleries. 7 caves are in this category: Be, Ct, Ko, Ma, No, Pr, Sg.
- Chambers. 18 caves belong to this category: Ar, Ba, Co, Ch, Fo, Ko, Ms, Mr, No, Or, Pi, Pe, Po, Ra, Ri, St, Sr.
- Collapse shafts. 4 caves belong to this category: Cm, Po, No, Ra, Ti.
- Porches. 2 caves are in this category: Bo, Re.

The term collapse has been reserved for caves where most of the ceiling has collapsed. Indeed, for certain cavities, the roof is crossed by a shaft whose dimensions are small, compared to the room. In this case, the stability of the vault is not compromised (Ar, Ba, Fo, Ms, Pe, Po). Thus, these caves have not been classified in this category.

Some caves belong to two categories. Kocain begins with a giant gallery and ends with a room. The entrance to the Aven Noir is a chamber whose roof is crossed by an aven (shaft); a giant gallery follows it and widens to form a chamber at its end.

Four chambers are junction rooms: Bo, Ch, Co and Pi.

- Bournillon (Bo): The porch connects the wet cave network with an upper dry gallery, the Salle Rose (Pink Hall).
- Champclos (Ch): The room is located on the path of an underground river. From each corner starts a fossil gallery. Active water circulations are joined in the north and southwest galleries. The southeast gallery is an upper dry gallery. The hall thus occupies a junction position which seems to have favored its genesis.
- Coufin (Co): The chamber is located at the junction between two underground rivers.
- The Verna Room is a special case; it can be considered as a vertical junction between the upper galleries (Amont and Aranzadi galleries) and the conduits of Arphidia Cave located beneath.

3.3 Geographical Location

3.3.1 Regional Distribution

The cavities studied, with the exception of one, are located in the five following regions (Fig. 3.1):

- Pyrenees (Be, Ma, Pi, Pe, Pr, Ri, Rt, Sg, Sr),
- Cevennes (Ar, Ch, Fo, No, Or),
- Quercy (Pa, Re, Ti),
- Vercors and Chartreuse (Bo, Co, Ms, Mr) and
- Jura (Ba, Po, Ro).

The other identified but not studied large caves are also located in these same areas.

One cave only is in Provence (Castelette) although the karst phenomena are numerous in this region. It is a small volume chamber. Its ceiling height does not exceed 10 m. The largest volumes are in the Pyrenees. Chambers and giant galleries are much more numerous than in other regions. The importance of rainfall and the presence of thick limestone series at high altitudes appear to play a role in this concentration (Fig. 3.2). There are even large rooms (50 and 35 m wide) at 2,820 m.a.s.l. in the Devaux Cave on the edge of the Gavarnie Cirque. The large volumes, therefore, seem to be linked to the existence of a significant karst potential resulting from the combination of several lithological, climatic and structural parameters:

- thick limestone massifs,
- abundant rainfall and
- rapid waterflow, thanks to steep slopes.

Referring to the current climate, the preferred areas for finding karst large volumes are the mountains and the tropical zones. For the latter, a very important rainfall can compensate a gentle topography as for the world's largest example: Sarawak Chamber in Malaysia.

The age of the caves is, however, poorly known; it is therefore not possible to specify all the climatic phases that accompanied the genesis of the cavity. It is certain that a high rainfall is a favorable factor for digging by dissolution, erosion and evacuation of the material. Indeed, whatever the predominant genesis factor, it is also necessary to prevent the caves from being clogged by clay or filled by speleothems. On the other hand, arid regions may contain large but clogged caves.

3.3.2 Relationships with the Surface

3.3.2.1 Thickness of Rock Above the Ceiling

The depth is currently over 1,000 m in many cave networks. During the thesis work in 1985, the world record was in Jean Bernard Shaft (−1535 m). In 2019, it is Veryovkina Cave in Georgia (−2212 m). However, the depth of a cave is the difference of elevation between the highest entrance and the lowest part. It does not reflect the thickness of the limestone coverage. The deep caves extend more or less near the slopes

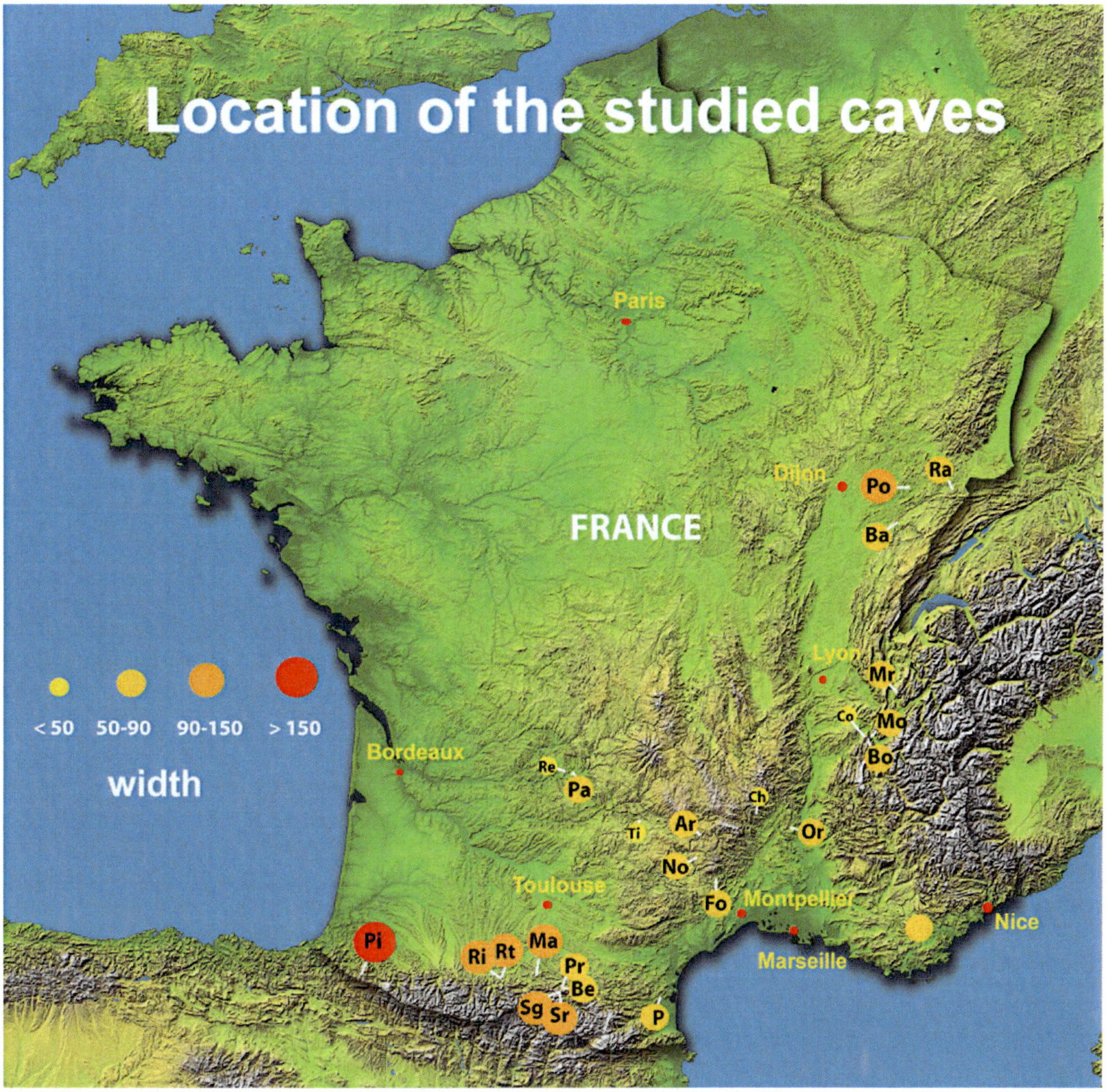

Fig. 3.1 Location map of the studied examples

from which they deviate at maximum 500 m (Berger Chasm, Gouffre Jean Bernard Shaft, Pierre St-Martin Shaft, etc.; Renault et al. 1967; Gèze 1965).

This observation has various explanations. Some authors consider that beyond this distance the discontinuities are not open enough to allow water circulation and thus karstification. It should be noted, however, that the limestone series do not exceed 500 m thick in France. These data should be compared with what is known abroad. Examples in Papua New Guinea, where the series are very thick, nevertheless, seem to confirm that. Among the studied caves, only three have a limestone cover that exceeds 100 m:

- Bournillon porch (350 m),
- Coufin (260 m) and
- Bedeilhac Cave (280 m).

The first two caves are near the slope. Bedeilhac Cave is, however, the only example really located in the heart of a massif, under nearly 300 m of limestone. In all three cases, the rock is a limestone with an Urgonian facies. Except for these three caves, the thickness of rock above the chamber roof is lower than their span and varies between 15 and 100 m. The former value seems to be a minimal one which requires that the limestone is not affected by a badly oriented cracking which would prohibit the stability of the ceiling.

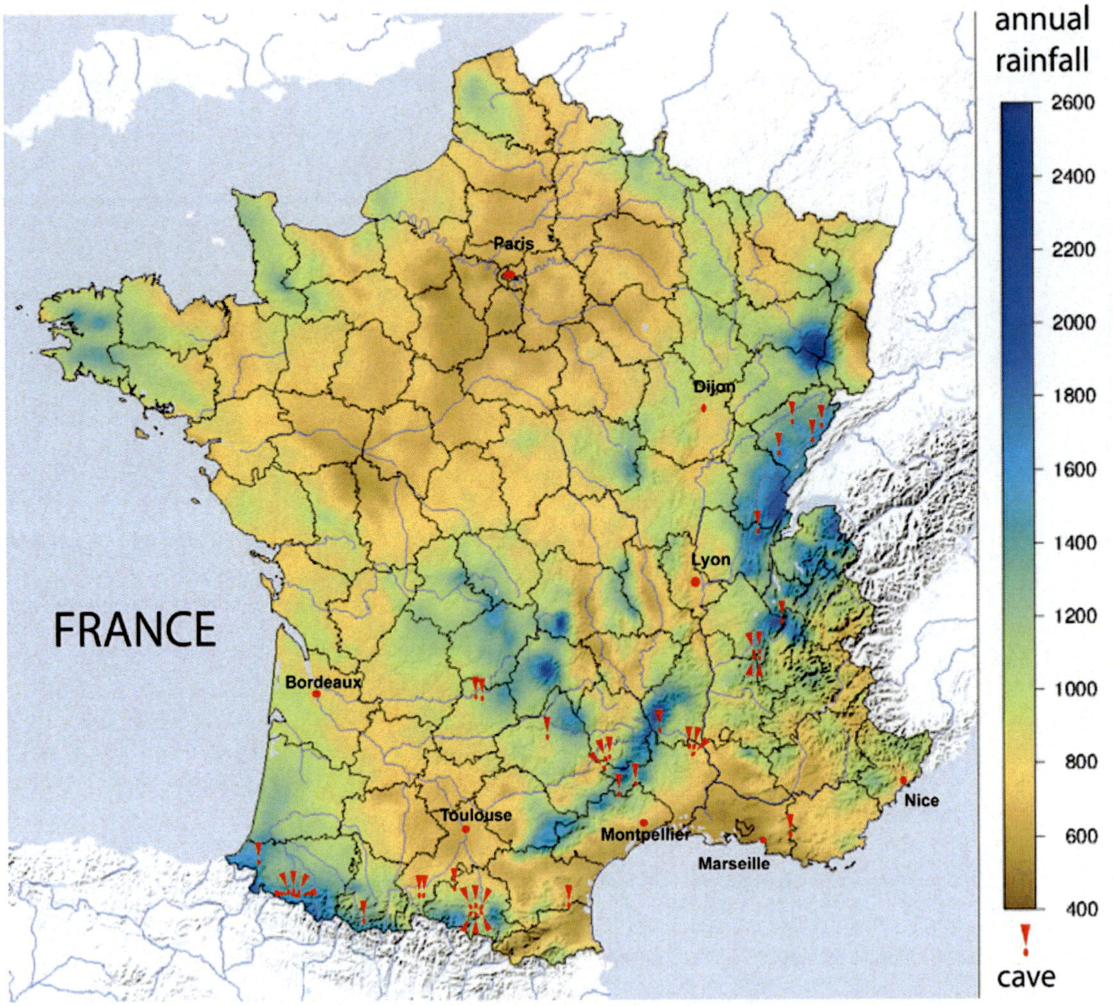

Fig. 3.2 Rainfall map. The caves are located in high pluviosity places

The rupture of the roof is the mechanism which gives birth to the collapse shaft (Cm, Pa, Ra, Ti). In such examples, it is sometimes possible to observe the partially preserved roof. The Ravières (Ra) or Hell (Cm) Shafts have vestiges of roof only 3 m thick; however, it is difficult to assess how important the limestone cover was during the chamber digging. Indeed, after the collapse the overlying land could be scoured by erosion, as well as the vault was able to continue to evolve in small successive collapses.

3.3.2.2 Proximity of the Slope

It seems that big underground volumes cannot exist deeply inside the limestone massifs; inversely, the proximity of the slope seems to favor the speleogenesis and even the oversizing of the hollows. Some examples seem indeed related to the proximity of a slope (Bo, Sg, Rt, Co, Mr). Thus, the Coufin and Chevaline networks widen suddenly and form chambers as they approach the Urgonian cliff (see Chap. 2.8, Fig. 2.56). However, one may wonder whether the current position of the cliff is not posterior to the widening of the rooms. Finally, without concluding, one can notice that the galleries are often oriented parallel to the slope (Renault 1968). For example, the axis of the Big Chamber of Sabart Cave (giant gallery) is parallel to the slope. These remarks lead to consider that the existence of a decompressed or tensile zone due to the incision of a valley augments the possibility for a karst river of excavating large volumes.

3.3.2.3 Relationships with Surface Morphology

In many cases, the large volumes are under a dry valley (Ch, Fo, Ma, Ms, No, Pa, Pi, Pe, Re). This fact could be related either to the need for large quantities of water to withdraw the excavated material or to a stress relaxation phenomenon. A particular case is that of La Verna Chamber located under Arphidia ravine which is an old glacial valley. At the vertical axis of the valley are located not only La Verna but also Styx Chamber (Lonné Peyré Shaft) and the labyrinths of Arphidia Cave which include several rooms. The ravine and the caves may have developed from a locally decompressed area. In this case, the caves may be anterior to the valley. Another explanation could be to consider that the digging of the glacial valley, then the melting of the glacier, caused a local decompression favoring the digging of the chambers. In that case, the ravine is older than the rooms.

3.4 Geological Context (Table 3.3)

3.4.1 Regional Context

Fourteen (14) caves are located in tabular series with slight dips (Causses, Cevennes, Vercors);
Five (5) caves develop within folded structures (Jura);
Ten (10) caves are localized in lands having undergone an important orogenesis with folds, thrust slices and thrust sheets (North Pyrenean border).

> La Verna Chamber is located in a particular geological context as the ceiling is hollowed out in the Cretaceous tabular series, which are laying in unconformity on the Paleozoic substratum where the lower half of the chamber is dug. Most caves are located near faults except Bournillon, Coufin, Orgnac and Poudrey.

Note: Except for Sarawak Chamber (Chap. 5), there is no cave located in an anticline structure.

3.4.2 Petrography

The nature of the host rocks is variable although sublithographic limestone is the most common (Ar, Ba, Be, Bo, Ko, Ma, No, Or, Pa, Pi, Pe, Po, Pr, Ra, Re, Si, Sg, Sr, Ti). There are also sandy limestones (Ba, Bo, Co, Mo, Mr, Pa, Re) dolomitic limestones or pure dolomites (Ca, Fo, No, Ri, Ti). Abroad, there are examples of large volumes in formations considered as weak like chalk (Australia) or gypsum (Eastern Europe). In fact, all the usual formations in which karst phenomena exist are present.

In several cases, the lower part of the cavity is hollowed out in soft materials that are usually not or hardly karstifiable:

- Marly limestone and marl: Ch, Po, Ra, Re, Ri, Rt.
- Schist and sandstone: Pi, Ri.

Note: These soft formations are not always visible in the cave; however, the geological study indicates their presence near the floor of the chamber (Ba, Bo, No).

When the cave is dug in several formations of different petrographic nature, the ceiling always develops in the formation that has the best mechanical resistance (e.g., sublithographic limestone or dolomite) (Ba, Bo, Ma, No, Pi, Po, Re, Ri, Rt).

3.4.3 Stratigraphy—Lithology

More than the petrographic nature of the rock, the thickness of the beds is an important parameter. Twenty-four cavities out of the 28 studied have ceilings hollowed out in formations layered in beds of more than 0.5 m thick. In some cases, the series is very massive, and the stratification is poorly or not visible (Be). The whole behaves like a single bed, tens of meters thick.

Surface alteration highlights the stratification joints; however, not all of these joints play a role in the terrain maintenance. The study of Moras Cave shows that the rock (sandy limestone) is formed by beds of about 5 cm. The joints are clearly visible at the level of the cliff over the entrance porch of the cave. Every meter, a less resistant layer is eroded and gives the rock an appearance of limestone in large metric beds. Inside the room, the centimetric bedding is not visible. The morphology of the vault is only determined by the metric spacing joints.

Inside Sabart Cave, the rock looks massive, without visible interbeds, while outside the observation shows that the rock is vertically stratified in layers 0.5 m thick. The Barrenc de Pla de Perillos presents a similar case. In the cave, the rock appears very massive; the stratification is visible only outside on the lapiaz surface. The stratification is generally much better highlighted outside than into the caves and many interbeds; especially, the dry ones do not play an important role in the stability of the rock. In many cases, the room is partly dug in series alternating hard materials (marl-limestone, sandstone, limestone) and soft ones (shale, marl) (Ch, Pi, Po, Ra, Ri, Rt).

3.4.4 Discontinuities

There are two kinds of discontinuities: sedimentary and tectonic.

3.4.4.1 Sedimentary Discontinuities

The interbed that emphasizes the rate of sedimentation forms a discontinuity between two limestone strata, particularly when the interbed is underlined by clay or marl. In that case, there is no material bridging between the beds. This is thus a surface without cohesion which may generate the collapse of the lower bed (Ba, Ch, Pa, Ti). The dry interbeds (without marly or clayey materials) are sometimes partially connected but can also be perfectly free (wall of Balsan Room in the Aven Noir Shaft, bases of the walls of the Bournillon Porch, Mort Ru Cave).

Finally, a special case is illustrated by the ceiling of the room of Castelette Cave. In that cave, tectonic movements caused bed slipping, which eliminated all cohesion between the strata as evidenced by the importance of the slickensides. The roof collapses gradually.

3.4.4.2 Faults and Joints

Most examples are located in highly fractured areas (Table 3.3). Inside the cave, it is not always possible to differentiate between fault and diaclase. The fracture walls are often covered with speleothems, or otherwise slightly eroded by peeling or dissolution. Sometimes, however, perfectly preserved fault mirrors are observed (Pe, Ri, Rt). This conservation may be related to the recent age of the collapse which cleared the wall or to a particular resistance of these sliding surfaces to erosion.

In general, the interbeds form predominantly horizontal discontinuity planes, which give them an important role in the structure and stability of the ceiling. Conversely, the faults and joints usually perpendicular to the stratification are predominantly vertical and mainly concern the walls. Among the studied cases, some cavities do not obey this rule. In Castelette Chamber, layer slipping is similar to faulting and concerns the ceiling. In Kocain, Riusec and Perillos, the joints are very inclined. In La Verna Chamber and in the Big Chamber of Sabart Cave, the bedding is vertical.

3.4.5 Role of Discontinuities

3.4.5.1 Morphology

The study of the large volumes shows that the wall or roof surfaces most often correspond to pre-existing discontinuities planes (interbeds, faults, joints), for example, Ar, Bo, Co, Ca, Ma, Ms, No, Or, Po, Ra, Re, where the ceilings are lower surfaces of limestone beds and Ar, Cm, Ko, Ms, Pe, Re, Ri, Rt, Ti, where the cave sides are fracture walls. The most spectacular examples are the flat roofs of Mas d'Azil Cave and Poudrey Shaft which are lower surfaces of bed planes and the walls of Riusec Cave and Tindoul de la Vayssière which are fault walls.

3.4.5.2 Genesis

The existence of large volume only caused by dissolution is a reality (Be); however, it is about exceptions located in very particular limestone. Indeed, there is still a large part of collapses in the genesis of large underground caves (see Chap. 5—Part 3). These collapses occur as soon as a block is dissociated from the massif by the discontinuities. It seems obvious that a fractured area is favorable to the genesis. A certain density of discontinuities is necessary for this digging to take place. In 19 caves, several sets of diaclases are observed. Only two examples do not show vertical fractures (Ch, Ca); however, on the other hand, horizontal discontinuities are very important there (alternations of marly limestone and marl beds: Ch; layer slips: Ca).

When the ceiling is cut by a tight network of large fractures, there is usually a collapse between these fractures (Ba, Cm, Fo, Or, Pa, Pe, Po, Ra, Ti) which forms a shaft. However, isolated joints frequently cut the ceilings without apparently affecting their stability (Ar, Be, Co, Ma, Or, Po, Sg).

3.4.5.3 Conclusion

The discontinuities thus play a dual role:

- They cut the limestone and allow the cave genesis by water circulation and dissolution in a first time, then by collapse.
- Their walls also form stable surfaces on which the enlargement of the cave stops.

3.5 Hydrogeology

3.5.1 Role of Water

In 11 of the studied caves, there is an active underground river (Ba, Bo, Ca, Ct, Ch, Co, Ma, Pa, Pi, Re, Ti). Some caves are traversed by rivulets unrelated to the digging and resulting from surface infiltration (Mr, No, Po, Re, Ri, Rt). The other large volumes are practically dry and appear stabilized in their digging. Nevertheless, it is not possible to imagine that cavities were dug without the help of large water circulation. Water is involved in two levels:

- on the one hand, the dissolution (limestone) or the mechanical erosion (marl, schist, etc.) of the rock.

– on the other hand, the clearing of the excavated material after the collapse of the roof and walls.

3.5.2 Ancient and Current Discharges

The most important discharges are those of the Arize and the Pierre Saint-Martin rivers. The Arize is a special case as the non-karstic catchment is very extensive compared to the volume of the cavity. In the Pierre Saint-Martin Shaft, the discharge is of the order of a cubic meter per second. The flow rates are significantly lower for the other caves.

Although it is difficult to estimate stream flows at the time when these volumes have widened, it may be thought that during the interglacial periods (cataglacial phases), the flows had to be larger (several cubic meters per second). For this reason, these interglacial periods are probably privileged ones for the genesis of large volumes.

3.6 Fillings

All the studied examples as well as most of the caves have a rocky floor masked by fillings. They testify to the mode of digging of the cave and its state of evolution.

3.6.1 Blockfields

3.6.1.1 Origin and Diversity of Blocks

The summary Table 3.2 shows that 25 caves have a floor covered with blocks, and 3 chambers (Be, Co, Pe) have a floor masked by a thick layer of clay that conceals the exact nature of the filling. This almost constant presence of angular blocks proves that all these large volume collapses play essential roles in the genesis of the cavity.

In all cases, some of the collapsed blocks are evacuated by the action of the underground river, thus creating the current volumes. The genesis of the large volumes could be achieved by a succession of these cycles of collapses and clearings. Castelette Chamber illustrates the situation after a collapse stage, because the filling still occupies most of the room and leaves only a meter high between the roof and the floor. On the other hand, the Mas d'Azil is at a different stage of evolution as most of the blocks were cleared by the river. This cave is, nevertheless, a special case; the discharge of the river is indeed important because it is fed by a large non-karstic watershed.

3.6.1.2 Size of the Blocks

The size of the blocks reflects the lithology (thickness of the strata) and the fracturing density. It is also often the only available information on the thickness of the limestone beds which form the ceiling of the cavity.

- small decimetric blocks of Mort Ru Cave where the rock is a thin bed sandy limestone (10 cm thick) and where the density of joints is important.
- huge decametric blocks in the Verna Chamber where the roof is dug in the massive Lower Cretaceous limestone.

There are sometimes huge unfractured blocks (60 m long in Trombe Hall in Cave, large blocks in Poudrey Shaft). Observations and logic lead to admit that these are blocks that have gently tilted under the effect of a scouring of their substratum and are not collapsed ones that would certainly have fractured during the fall. The blockfields are sometimes partially covered by finer scree. This is mainly observed near the entrances or at the bottom of the pitches. In some cases, these small blocks come from the ceiling and walls where underground weathering provokes rock spalling (Bo, Ms, Pa, Pe, Pr, etc.); in other cases, they come from outside the cavity (Ar, Ba, Or, Po, Ra, etc.).

3.6.2 Clayey Fillings

They are linked to the water circulation. They are deposits by settling of the clay contained in the water. Current and ancient fillings can be roughly distinguished.

3.6.2.1 Current Clay Fillings

They are linked with the presence of an underground river. The deposit is made only during floods. When downstream conduits clog up, the water forms a lake in the room, and clay settles on the blocks. This phenomenon was observed in the Baume Shaft. These recent deposits are observed in the following cavities:

- Baume Shaft at the base of the scree;
- Castelette Chamber at the base of the scree;
- Paradise Shaft (Cehennem) everywhere;
- Coufin Chamber, in all the rooms.

3.6.2.2 Old Clay Fillings

They are found in cavities currently abandoned by water (Be, Ko, Pe); the clay completely covers the soil, which is sometimes gullied by small superficial circulation. The absence of blocks on the clayey floor testifies the current stability of the ceiling in the three mentioned caves.

3.6.3 Allochthon Filling

In Sabart and Bedeilhac Caves, there are pebbles and gravels of various origins that are sometimes stuck on the roof or the walls. In A Hall (Riusec Cave), a gallery is totally blocked by morainal crystalline pebbles. These cavities were therefore partially or completely filled by these external deposits and then cleared later.

3.6.4 Speleothems

3.6.4.1 Repartition

The speleothem deposit is a good indicator of the state of stability of the caves. Some rooms are completely devoid (Mort Ru), while other ones are intensely decorated and sometimes the rock is totally masked. Stalagmitic massifs sometimes even completely clog a gallery. For instance, in spite of its large dimensions, the Demoiselles Cave was not studied during the thesis due to the importance of the speleothems that covered the ceiling and the walls which prohibited the study of the structure.

Caves that communicate widely with the exterior are almost devoid of concretions (Bo, Ma, Pa, Pr, Re, Ti). It is the same for cold caves (Pi, Ri, Rt). These observations confirm that the speleothem growth depends on the climatic conditions that prevail in the cave. In other cases, the absence of speleothems is obviously linked to the instability of the ceiling (Ca, Mr).

The most decorated caves are Ar, Fo, Ko, Or, Pe, Sg. Most of the speleothems are stalagmites and flowstone on the walls. Stalactites are less important; however, they often point out certain joints of the ceiling.

3.6.4.2 Supporting Role of Speleothems

The stalagmitic pillars, even when they come into contact with the ceiling, probably have no support role. On the other hand, the presence of flowstone on a wall can increase its behavior. This is particularly noticeable in the Mont Blanc Room (Fontaine de Champclos) where the marly levels are stabilized by a film of calcite.

3.6.5 Evolution of Filling

Blockfields are often partially cemented by flowstone. The speleothems may witness the instability and the evolution of the fillings. For example, in La Verna Chamber (Pi), below several large blocks one can observe an offset between stalagmites and stalactites that should normally be face to face (Fig. 3.3). This testifies that the blocks have moved. Their movement toward the bottom of the room (slope creeping) is provoked by a withdrawal phenomenon of the schists on which they lay. Similarly, the blocks that surround the funnel located at the bottom of the Poudrey Chamber have tilted. The movement is emphasized by the presence of inclined stalagmites (Fig. 3.4). To interpret these observations, two mechanisms may be involved:

- differential settlement of the clay under static loading (Be, Or) or drying of the clay,
- entrainment of clay by the water (racking).

3.6.6 Age of Speleothems

It is difficult to relate the age of a cavity to the importance or the aspect of the speleothems. In fact, only absolute dating makes it possible to judge. Thus, the stalagmites of the Aven d'Orgnac seem very old, given their size. However, for some, dating their base only gave an age of 6,500 years (Renault 1970). There are, however, in the scree, large stalagmites that are certainly much older.

Note: Since the thesis work dating was performed, ages can be older than 450 ky (Delannoy and Jaillet 2007).

3.7 Cave Shape

3.7.1 Conditions of Observation

The fillings are present in all the caves. They mask the lower part of the rooms and do not make it possible to measure the real volume affected by the digging. On the other hand, the vaults are generally widely visible, and the speleothems very rarely mask the morphology. In many cases, the ceilings are more or less arched vaults which lower to the floor without morphological discontinuity. One cannot distinguish the limit between the ceiling and the walls (Be, Ch, Ko, Pe, Pr, Sg).

In other cases, vertical walls are observed. The cave of Mort Ru is an example where a rounded ceiling arches over vertical walls. The shape of the connection between vault and walls indicates the current genesis of this cavity. The walls develop here in decompressed rock surmounted by a natural vault. These walls are gradually sinking toward the center of the cave. This evolution tends to form a shape that is observed in the northern part of the room where the decompressed vertical walls have fallen, revealing the lower parts of the vault (Fig. 3.5).

In Mort Ru Cave, the current vertical walls thus appear as a transitory morphology. On the other hand, some examples show the presence of stable vertical walls. They are often constituted by the wall of a major fracture and appear stable (Ar, Ba, Ct, Ms, Or, Pi, Pe, Ri, Rt, Ti).

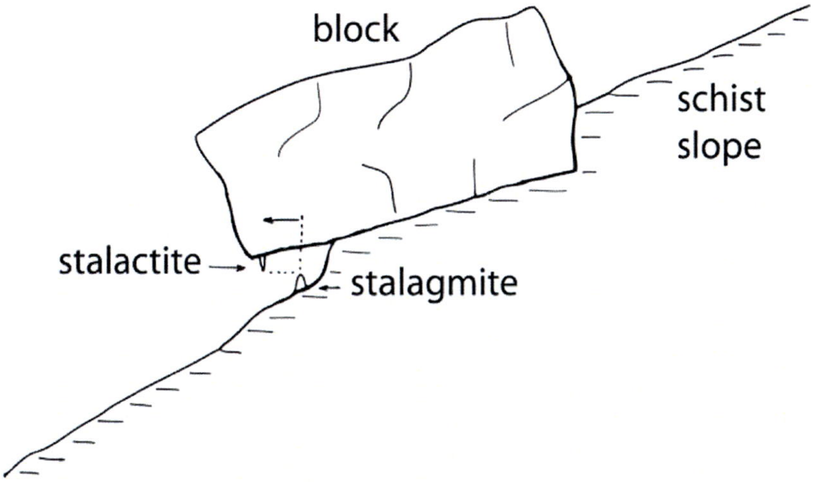

Fig. 3.3 Speleothems offset provoked by the displacement of a block toward the bottom of La Verna Chamber

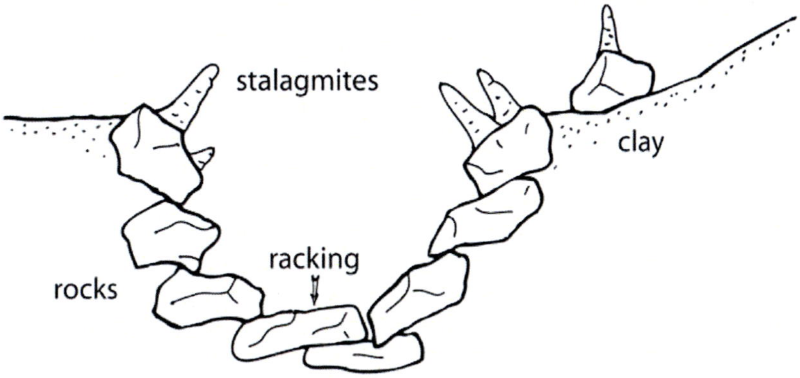

Fig. 3.4 Suffosion funnel in Poudrey Shaft

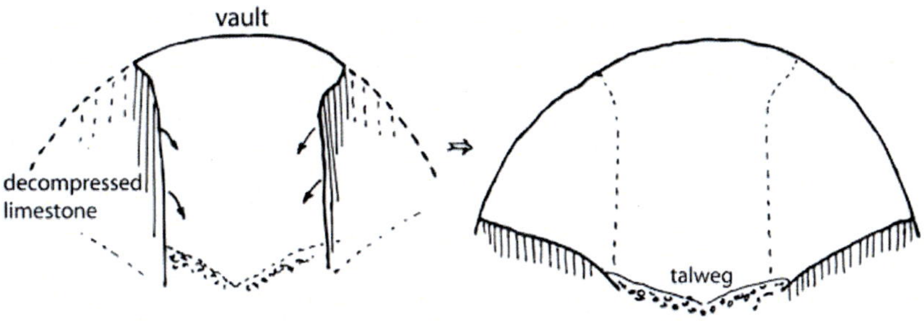

Fig. 3.5 Evolution of walls in Mort Ru Cave

Note: The evolution described above creates a talweg-shaped filling without involving a water circulation or a withdrawal. In fact, the cave floor receives more blocks near the walls than at the center of the cavity which gives a V-shaped section to the filling (see Sect. 2.13.2 and Fig. 2.94).

3.7.2 Various Shapes of Ceiling

There are two types:

- rounded or domed vaults,
- structural ceilings.

3.7.2.1 Rounded or Domed Vaults

It is the shape which theoretically ensures the best distribution of stresses around the cavity; however, the anisotropies of the limestone massif: bedding, joint set and faulting, often cause the birth of an asymmetrical vault (Ko, Pe). When the anisotropy is well marked, the stress distribution field is probably also deformed. The vaults can be more or less arched from slight rise arches (Be) to very rounded ones (Pi, Ch).

3.7.2.2 Structural Roofs

The shape of the ceiling reflects the structure of the limestone mass in which the cave is dug. The surfaces (walls and roof) that limit the cave then correspond to the discontinuity planes of the rock. They are either the lower plane of a limestone bed (e.g., large slab of Poudrey Shaft, Mas d'Azil Cave or Noir Shaft) or the wall of a big fracture (e.g., southern part of the ceiling of Trombe Hall in Riusec Cave).

A very common case is the corbelled vault or "reversed staircase" (Figs. 2.1 and 3.6). It is an intermediate form between the slab type and the rounded arch type. Each limestone bed protrudes inside the cave, with a more or less large span, and overpasses the underneath bed. The keystone of the ceiling is then formed by a slab that can reach a large span (Ar, Ba, Bo, Ma, Ms, Or, Ra, Re). The span of each bed depends on the cohesion of the limestone and on its fracturing. This type of ceiling is mainly observed in the sublithographic limestone massifs, when the layers have a thickness greater than 0.5 m.

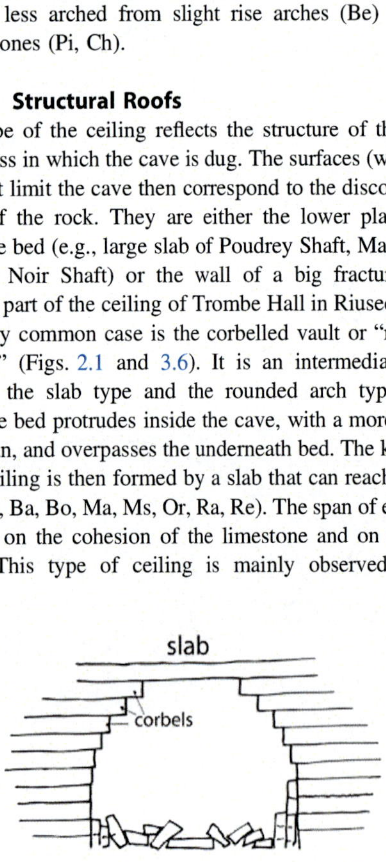

Fig. 3.6 Corbelled ceiling

3.7.3 Comparison Between the Observed Forms

3.7.3.1 Rounded Arches

The rounded arch shape is a mechanical equilibrium shape. It is observed in ancient caves (Be, Ko, Pe, Pr, Sg, Sr) or in rocks with a weak cohesion (alternation of marls and marly limestones of the Mt Blanc chamber in Champclos Cave, limestone in very fractured small beds of Mort Ru Cave). In these cases, the ceiling is a compression form.

The vault is an "active" form. The materials that were under the vault collapsed gradually. The equilibrium is quickly reached in inconsistent rocks. It will settle much more slowly in resistant rocks. In the latter case, the ceiling remains stable without the rounded arch shape being reached (Fig. 3.7). This concerns the corbelled vaults. The limestone beds have a sufficient mechanical resistance to tensile stresses that makes possible the existence of overhangs. The thickness of the beds and the density of the fracturing are thus factors of first importance which determine the roof morphology. However, it may be questioned whether these corbels are formed of rock elements situated below an active vault and thus only supporting themselves, or if each overhang supports a part of the overlying weight like a beam. The observation of openings between the beds would eliminate the second hypothesis.

3.7.3.2 Slab Roofs

When the ceiling is constituted by a large slab, the same problem arises. Does the bed that forms the ceiling participate in the holding of the vault like a beam or does it only support its own weight? (Fig. 3.8).

3.8 Chamber Genesis

3.8.1 Digging Factors

The omnipresence and the importance of blockfields in the studied caves prove the important role played by the collapses for the digging of the large underground volumes. The karstification starts initially by dissolution but the presence of a large volume requires a phase of collapse thereafter. Only one example seems to escape this rule (Be). Whatever the mode of digging envisaged, it can be considered that a large volume is established when there is a possibility of clearing a decompressed material below a natural stable vault.

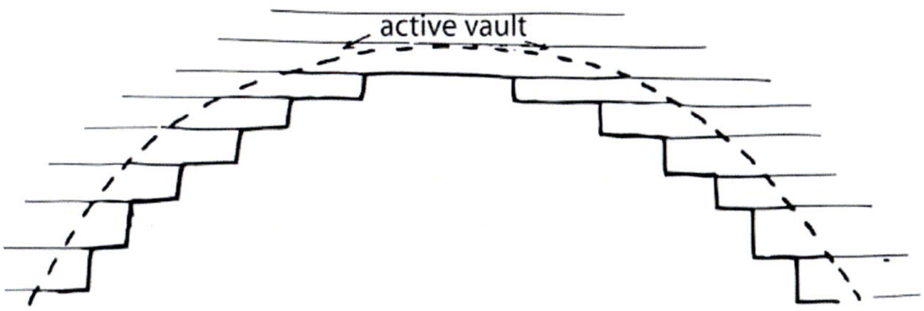

Fig. 3.7 Corbelled ceiling below an active vault

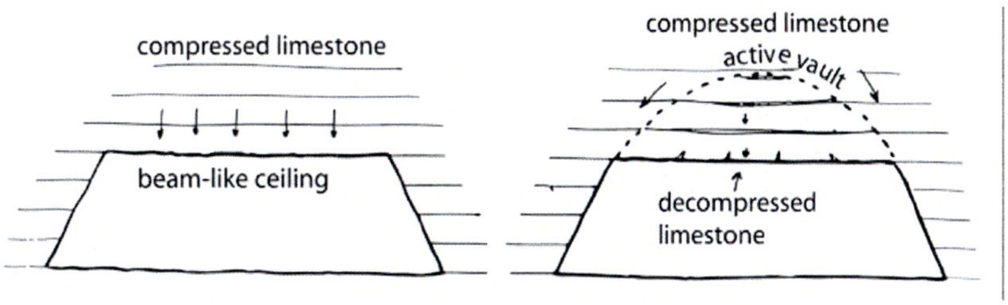

Fig. 3.8 Slab-roof: beam-like slab (left) or slab below an active vault (right)

3.8.2 Different Types of Digging

3.8.2.1 Dissolution (and Collapse-Dissolution) (Be)

The gallery is dug by the water that dissolves the limestone. The cracks are widened; the blocks are delimited by these cracks collapse and are dissolved by the river. This implies a digging in the saturated zone of the karst aquifer.

Bedeilhac is the only example but the presence of a thick layer of clay on the floor does not make it possible to observe the exact nature of the filling.

3.8.2.2 Dissolution and Collapse

This is the most common type. The widening of a karstic conduit by dissolution in the vadose zone creates overhangs that cause rock collapses. The collapsed material is then dissolved by the river or mechanically driven. It is the graviclastic type described by Montoriol Pous in Renault (1957, 1969).

3.8.2.3 Scouring-Racking and Collapse

The underground riverbed widens when reaching a soft formation (clay, marls, schist). The enlargement creates overhangs that evolve by successive collapses:

- marls in Ch, Po, Ra (and probably Ba, No);
- clayey-limestones in Ma, Re;
- schists in Ri, Rt.

This implies the existence of a mechanical clearing of the insoluble material by the river, toward downstream. A special case is that of racking. The soft material is drained down vertically, thanks to the existence of a lower network of karst conduits (Pi, Po).

Note: The presence of a soft material under the limestone is perhaps also a favorable factor if this allows a better stress distribution by deformation of the lithologic contact.

3.8.3 Evolution

In most cases, the cave then tends to reach an equilibrium shape which is most often a rounded vault. This equilibrium state can be broken when the enlargement continues or when the constraints vary within the massif, either because other caves are dug or because the slopes evolve by vertical erosion of the nearby valleys. It can also be caused by disappearance of a glacial overload.

3.8.4 Current Stability of Large Underground Volumes

All the studied chambers are stable. There are unstable zones; however, these are local phenomena that do not affect the stability of the entire cave.

The collapse shafts (Cm, Ct, Pa, Ti) are examples of collapsed chambers. They seem to have been created long ago and most likely during climatic crises (glacier melting, for example). They are related to an unfavorable rock structure (marly levels, fault bundle, etc.).

Paleogeographic studies show that the main lines of the current topography were acquired during Early Pliocene. The modifications are mainly related to the Quaternary glaciations. These glaciations had a dual role: on the one hand, a water storage and on the other hand, marine regressions and a slowdown of the water flows.

During post-glacial phases, large quantities of water could be mobilized with a significant hydraulic potential that favored the incision of the valleys. The result in the karst massifs was then a change of the groundwater base levels and a variation of the stresses related to the deepening of the valleys (Renault 1970). These variations were rapid (a few thousand years).

The major steps of large volumes digging probably correspond to these post-glacial periods. The stress variations cause collapses, and the large quantities of water favor the clearing. After a "fast" phase during which most of the volume has formed, the cavity evolves slowly until reaching a high stability. All the studied caves are in the adjustment phase (Mort Ru) or are totally stable.

The digging of the karst caves seems to be a succession of equilibrium stages separated by rapid states of instability. This hypothesis is supported by the fact that many large karst networks are organized in well-individualized levels connected to each other by pitch zones. Each level corresponds to a phase of equilibrium with the groundwater base level.

3.9 Conclusion of the Fieldwork

The genesis of the karst underground large volumes can result from a classical mechanism of karstification: the dissolution of a soluble rock (limestone, gypsum, etc.) that can also be a mixed process of collapse then dissolution. However, another way of digging, the scouring/racking process, must be taken into account as it explains the genesis of the largest volumes in the world, including the Sarawak Chamber in Borneo. This digging process involves the mechanical withdrawal, by karst groundwater, of an insoluble but erodable rock located below a thick limestone roof.

This raises the question of whether large underground volumes may exist in non-carbonated rocks. The existence of immense sinkholes in the quartzites of the Sarisarinama Plateaus (Venezuela) could provide a positive answer. A classical dissolution process is doubtful due to the low solubility of quartzites. Thus, the existence of an erodible rock level beneath the quartzite should be verified.

Unfortunately, the data on similar caves are scare because the non-carbonated geological environments are usually considered as unfavorable and are neglected by speleologists. Whatever the digging mode envisaged, water is a clearing agent either by dissolution or by mechanical transportation. The relative importance of both modes of evacuation remains to be clarified.

The chambers that have been explored are stable, which proves that the presence of joints does not exclude the existence of underground large volumes. Indeed, the width of 50 m, which is the lower limit of the studied caves, makes that most of them develop into a rock volume that includes numerous lithological and tectonic discontinuities. The existence of large volumes is only possible if an arrangement of the rock stress around the chamber prevents the collapse of blocks framed by the discontinuities network. This arrangement creates, by compression, an active arch around the void. The observed forms perfectly illustrate the theoretical results of rock mechanics (Goguel 1953).

The active natural vaults probably exist before the cavity is created. The large volumes then correspond to a removal, by the water, of the decompressed rocks located beneath this vault. The large volumes could, therefore, be prefigured inside the massif. The mechanical properties of the decompressed rocks located below this vault define the shape of the ceiling. If these materials are very resistant (very massive limestone beds), the ceiling may be slab-shaped (Poudrey). Stress measurements in boreholes through the roof could allow the highlighting of this active vault and the underlying decompressed zone.

References

Delannoy, J.J., Gauchon, C., Jaillet, S.: L'Aven d'Orgnac, valorisation touristique, apports scientifiques. Collection Edytem, Cahiers de Géographie n° 5, 178 p (2007)

Gèze, B.: La Spéléologie scientifique. Le Rayon de la Science, Seuil edit., Paris, Coll (1965)

Goguel, J.: Données techniques sur l'effondrement des cavités souterraines, VIII-1, pp. 1–8. Ann. Spéléo. CNRS edit, Paris (1953)

Renault, P.: Sur deux processus d'effondrement karstique, XII-1–4, pp. 19–46. Ann. Spéléo. CNRS, Paris (1957)

Renault, P.: Influence des pressions de terrain sur la génèse des réseaux de cavernes. 3ème Congrès Suisse de Spéléo., Interlaken - La-Chaux-de-Fonds (1967)

Renault, P.: Contribution à l'étude des actions mécaniques et sédimentologiques dans la spéléogénèse, pp. 1–23. Ann. Spéléo. CNRS, Paris (1968)

Renault, P.: Sur les phénomènes de détente mécanique qui sont à l'origine des réseaux de cavernes dans le sous-sol des karsts. C.R Acad. Sc. Paris **268**, 1168–1170 (1969)

Renault, P.: La formation des cavernes, 129 pp. Coll. Que sais-je? n° 1400, P.U.F. edit., Paris (1970)

4 Application to the Digging of Artificial Chambers

This chapter presents the artificial underground chambers and the methods used for their digging in the 1980s. Then, the geological environments of the natural chambers studied during the fieldwork are analyzed to define favorable environments for creating very large artificial chambers. The idea of searching a natural vault above soft material is advanced, and several types are presented that are suggested by natural examples. The emptying of the marly heart of a limestone anticline could be a satisfactory solution.

4.1 Classical Artificial Chambers

4.1.1 Various Types of Underground Realizations

The modern world is more and more interested in the construction of underground works. The lacks of urban floor space, aesthetic, strategic or economic purposes are the main causes. This is why EDF (Electricité de France) operates currently 34 underground hydroelectric power plants. The plants require large volumes for the installation of the equipment. They must be placed as close as possible to the reservoir and have large section penstocks (10–100 m^2). These penstocks can then be dug directly into the rock instead of building steel pipes. There are more than 300 underground power stations in the world, including 150 in Norway.

In carbonate rocks, limestone or dolomite, these achievements are rare; the dimensions of the engine rooms are in the following range: 12–33 m wide, 22–41 m in height and 35–145 m in length. In France, the largest span is that of the Sautet plant with 36.5 m width before coating. This artificial chamber is dug in limestone (Rescher 1975).

Hydropower plants are not the only major underground projects; for example, the Rove tunnel linking the Berre pond to Marseille harbor. This tunnel, which is currently partly collapsed, has a span of 22 m and extends in limestone series (Bernède 1972; Duffaut 1982).

The cavities dug in salt, by dissolution, reach impressive dimensions (Obenaueur et al. 1982). West Hackberry (USA) is 135 m in diameter and 290 m in height. Manosque (France) is 87 m in diameter and 260 m in height. These underground volumes have particular shapes with a large vertical development that favors the stability of the ceiling. They are never empty (hydrocarbon, water) which contributes to the stability of the walls. These cavities, nevertheless, undergo deformations and sometimes they collapse (e.g., the experimental cavity of Kiel) (Berest 1981).

The largest underground artificial volumes, dug in limestones without pillars, are in Finland in Tytyri quarries. Operating chambers, in Solhem sector, measure up to 100 m × 100 m × 100 m. The digging of these chambers necessitated the continuous measurement of the deformations of the walls to adapt the shape of the cavity to the stresses. In this case, the stability of the roof is favored by the existence of important natural horizontal compression constraints (Hintikka 1969). This example shows the interest of adapting the chamber shape to the stress field of the rock massif (Duffaut 1977). This is not often possible, especially when the room must house specific equipment, as is the case for the hydroelectric plants. The keyhole shape is generally better suited to welcome the hardware.

4.1.2 Mode of Realization of Underground Hydroelectric Power Plants (After Plichon et al. 1978)

4.1.2.1 Location

The location is generally imposed by the pattern of the fall within a hundred meters. It is most often along the river.

E. Gilli, *Big Karst Chambers*, Advances in Karst Science,
https://doi.org/10.1007/978-3-030-58732-1_4

Fig. 4.1 Cross section of the man-dug Oymapinar caverns

Geological and geotechnical reconnaissance aims to find the best possible location. It is preferable, as far as possible, to place the plant in cross-cut position. Finally, acceptable solutions are few.

4.1.2.2 Studies and Recognitions

A recognition gallery allows to study precisely the geological context and to collect rock samples on which tests will be made:

- shear strength,
- tensile strength,
- elasticity module,
- cohesion and internal friction angle of the joints,
- density,
- porosity and
- hydraulic tests.

Flat-jack tests indicate the stress field state.

4.1.2.3 Plant Shapes

The keyhole shape is the best one as it helps to welcome the equipment. On the other hand, the risk of buckling of the walls augments because the horizontal stresses are important. Thus, when the rock has bad mechanical properties, the following shapes are used:

- horseshoe: Saussaz II, Montezic (France);
- ovoid: Waldeck (Germany), Darguinah (Algeria);
- circular: Batang Padang (Malaysia).

Note: When excavating an underground hydroelectric plant, the engine room is not the only underground construction; it is necessary to dig other cavities that change the stress field and augment the decompressed areas. One example is the Oymapinar plant in Turkey, dug on a cliffside in dolomitic limestone. There are two chambers, 105 m long and more than 20 m span, that are separated by 20 m of limestone rock (Horn 1982) (Figs. 4.1 and 4.2).

4.1.2.4 Digging Phases

The excavation begins with the ceiling. Longitudinal galleries allow to reach the desired profile, in several steps, while increasing the stability of the rock with tie rods. A concrete shield is progressively placed on the roof. When

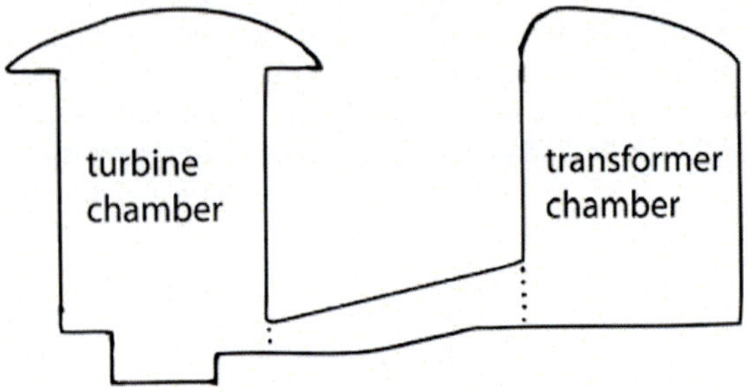

Fig. 4.2 The digging of Oymapinar caverns

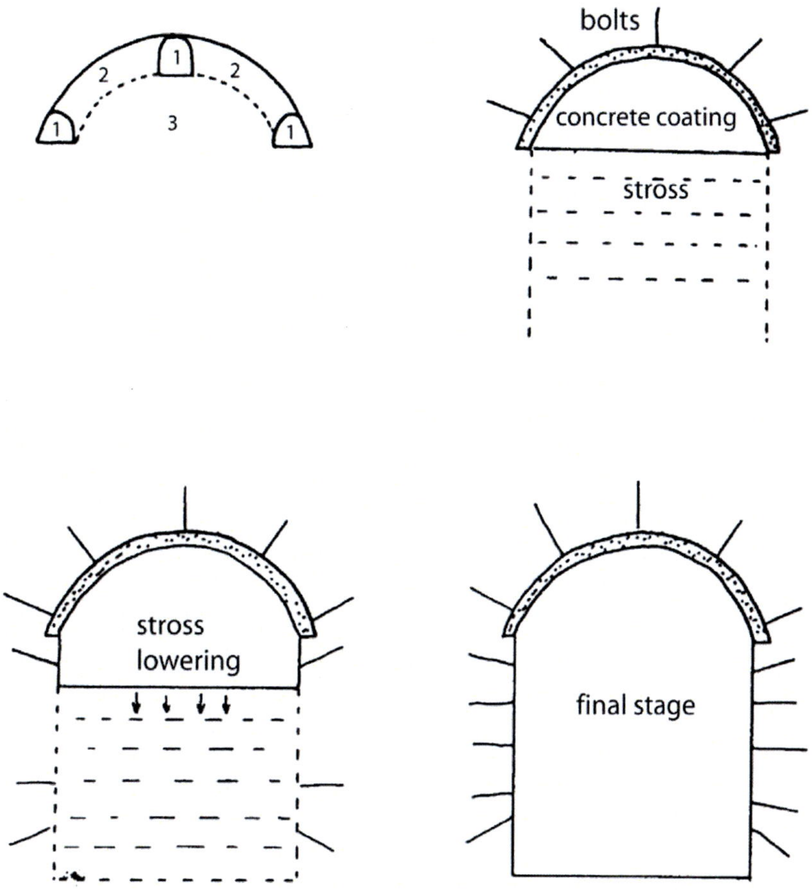

Fig. 4.3 Different digging stages of man-dug caverns

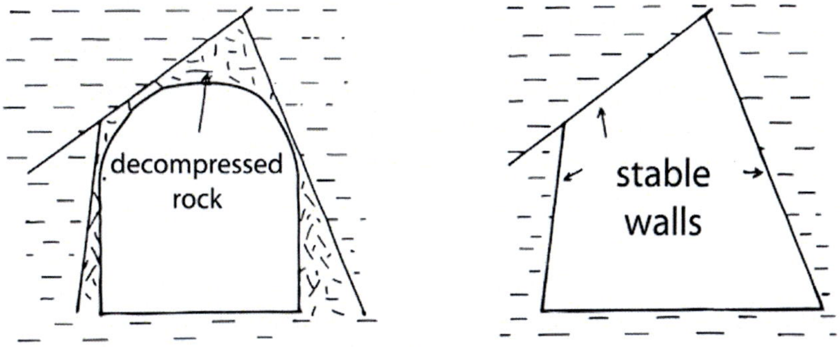

Fig. 4.4 Use of stable fracture walls

the rock quality is good, it has only a role of protection and sealing; and it does not support the roof. Once the final vault is completed, the excavation starts downward till the floor level. Tie rods are also installed in the walls according to the estimated deformations (Fig. 4.3).

4.2 Application of the Thesis to the Creation of Very Large Artificial Chambers

4.2.1 General Principles

The inventory of large underground volumes shows that a 450 m span exists in Borneo. Four French examples are greater than 100 m. The limit of 50 m is largely exceeded. In all the studied cases, the dimensions are greater than the spacing of the natural discontinuities (interbeds and fractures). Their presence is not a crippling factor. However, the examples of collapse shafts show that when the density of a vertical fracture is too high the roof breaks (Ba, Cm, Dt, Fo, Pa, Ti). These are always examples located near the surface.

Natural cavities exist in decompressed areas located below a natural stable arch. The role of water consists in clearing the decompressed material. A similar process could be considered for digging artificial cavities by

- searching a natural vault within the massif;
- digging the chamber under this vault, in a soft material.

4.2.2 Remarks on the Stability of Fracture Walls

The study shows that fracture walls are stable surfaces. It, therefore, seems interesting to use them as natural walls in an artificial chamber dug in a fractured environment. Rather than giving the cavity a specific shape (keyhole, for example), the digging could create a volume determined by the geometry of the fractures. It may be more cost-effective to remove decompressed material under a fracture than to reinforce it with tie rods (Fig. 4.4).

4.2.3 Favorable Structures

Diverse types are suggested by natural examples:

- The Poudrey type (Fig. 4.5): The ceiling is formed by a thick limestone bed. The volume is excavated in the underlying marls. The walls are formed by embankments shaped in the marls, coated or not. In plan view, a circular shape allows a greater stability by horizontal arching effect.
- La Verna type (Fig. 4.6): If the resistance of the overlying limestone is insufficient, or if the span is very large (>100 m), the roof can be dug to form a rounded arch.
- Anticline type (Fig. 4.7): This leads to consider the anticlinal type, where the arch is then "geological" (Duffaut 1982). This consists in emptying the marly heart

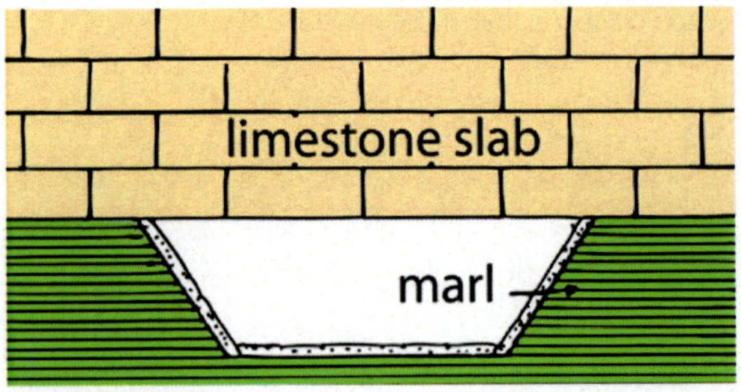

Fig. 4.5 Poudrey type

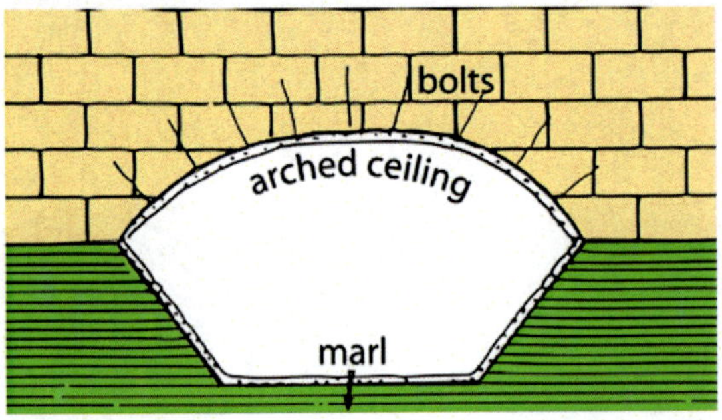

Fig. 4.6 Verna type

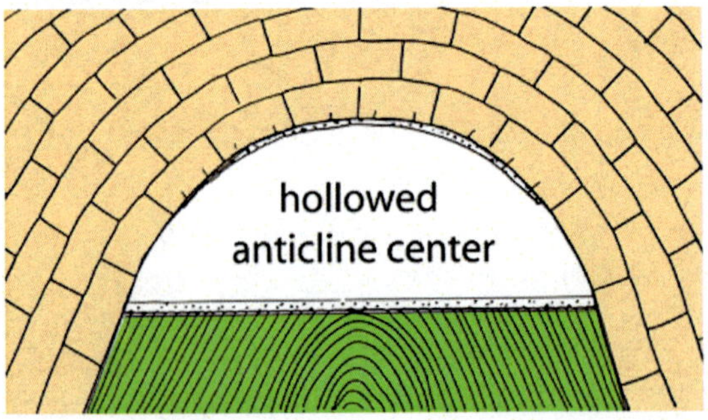

Fig. 4.7 Anticline type

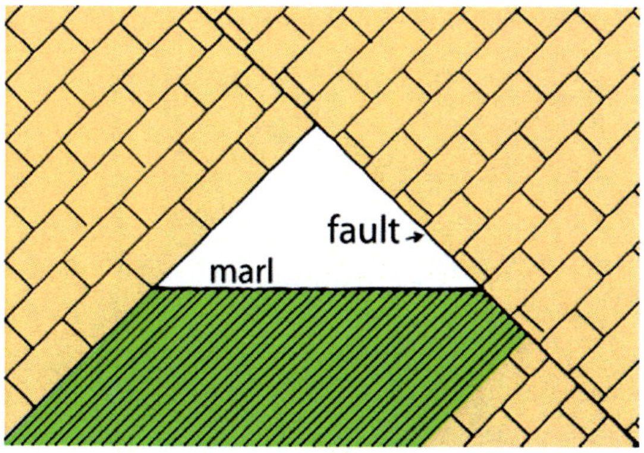

Fig. 4.8 Fault type

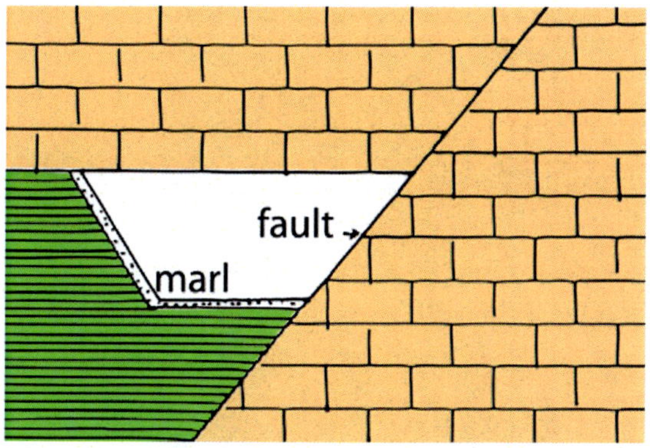

Fig. 4.9 Another fault type

Fig. 4.10 Riusec type

of a limestone anticline. This example is close to what is observed in the north corner of the Trombe Hall in Riusec Cave. Natural examples are rare because anticlines are structures that are generally unfavorable to the development of a karst network. In fact, groundwater circulation is preferably located in monocline structures or syncline gutters. Anticline types would make possible the digging of great span chambers. Unfortunately, such environments are rare in a given area.

- Fault types (Figs. 4.8 and 4.9): Many examples show the role played by the large fractures whose walls constitute a stable plane. These caves also appear to have great stability. Artificial cavities could thus benefit from the use of a fault wall to form a stable wall or a portion of ceiling. In "Riusec type", a fault forms the wall and the roof of half of the room (Fig. 4.10).

4.3 A Page of Humor

The thesis on underground volumes ended with a humorous note (Fig. 4.11) from the famous French comics author J. M. Reiser (1941–1983).

Fig. 4.11 A Reiser's comic that concerns underground chambers

References

Berest, P.: Stabilité des cavités de stockage d'hydrocarbure dans le sel. Données de l'expérience internationale. Revue Française de Géotechnique **16**, 5–10 (1981)

Bernède, J.: Le tunnel du Rove. T.E.G.G bull, EDF, 13185-23/12/1981 (1972)

Duffaut, P.: Quand et comment convient-il de dimensionner les structures souterraines? Journées d'études "Les travaux souterrains en site urbain", Paris, May 1977, AFTES, 135 (1977)

Duffaut, P.: L'effet d'arc en souterrain et ses limites. Séminaire Franco-Suisse géotechnique des tunnels. Lausanne (1982)

Hintikka, J.: Spécial discussion on large. In: Permanent Underground Openings Symposium, Oslo, p. 271 (1969)

Horn, K.: Construction des cavernes et des puits sous pression de la centrale hydro-électrique d'Oymapinar en Turquie. In: ISRM Symposium Aachen 1982/05/26-28, p. 293 (1982)

Obenaueur, P.W., Lielups, L., Ruse, P.: Creusement et examen de la stabilité des grandes cavernes (artificielles). In: I.S.R.M. Symposium Aachen 26-28/05/1982, p. 397 (1982)

Plichon, J.N., Le May, Y., Gomes, G.: Centrales hydroélectriques souterraines. Réalisations récentes d'EDF. Tunnels et ouvrages souterrains. Bull. A.F.T.E.S. Paris, 30

Rescher, O.J.: Réalisations récentes de grandes cavernes pour usines hydroélectriques. Bull. A.F.T.E.S. Paris (8), 60 (1975)

5 Study of Sarawak Chamber

This chapter presents the world's largest example of an underground karst chamber: Sarawak Chamber in Borneo. It was only shortly described during the thesis work (Chaps. 1, 2, 3 and 4), and a specific project took place in 1993. The Chamber is located in the Mulu Range where hundreds of kilometers of caves were explored. In addition to Sarawak Chamber, Deer Cave, a very large gallery, is also presented. Both caves are located at the contact between a thick unit of limestone and a substratum of clayey sandstone which confirms the conclusion of the thesis work on large underground volumes (Chaps. 1, 2 and 3) and the hypothesis of big chambers located below an anticline (Chap. 4).

5.1 Origin of the Project

During the literature search for thesis, data were found the discovery in 1980 of Sarawak Chamber, the largest example in the world. The author wondered if it was in the same environment as the largest French examples. The following text had already been written on the chamber as an appendix to the thesis work, but it was succinct.

5.2 Short Description of Sarawak Chamber in the Thesis Work

.... This chamber, the world's largest one, belongs to Lubang Nasib Bagus cave which is located on the course of an underground river that crosses a limestone massif, the Melinau formation. It is reef limestone unit dated Eocene. The average dip of the series is about 45° to the west. In the east the Melinau formation lays on the Mulu formation, a thick series of sandstones and shales. Above the limestones, the sedimentary series continues with another impermeable set, the Setap shales. The water flowing on the impermeable terrains of Mulu Formation sinks underground when arriving in contact with the limestone. The resurgence is in the southwest, at the contact between the limestone and the Setap shales.

The Sarawak Chamber is located at the contact between the Mulu formation and the Melinau limestones. The river flows on the impermeable shales and sandstone of Mulu formation. The riverbed widens by scouring this formation which caused the collapse of the ceiling. It is thus a phenomenon of racking similar to those of La Verna chamber and both Riusec halls (Fig. 5.1)...

5.3 Sarawak 1993 Project

Thus, after the 1984 thesis, I tried to find financial support to study Sarawak Chamber with the caving group as the CEK (Karst Study Center) specialized in the study of large volumes. The initial project was a comprehensive study of the chamber; however, the lack of credit only led to a simple reconnaissance in 1985, followed in 1993 by a light photography mission with a limited budget. The cavers decided to simply make a stereo photo and to collect information on the geological context of the main large volumes. In order to develop the techniques of shooting, many training sessions had been carried out in French Caves (La Verna Chamber and Poudrey Shaft) and also in limestone quarries (SPADA, Nice). The expedition, with Pierre Delange, Thomas Gaschat and Remy Schejbal took place in March 1993. It was successful. Five days were spent in the chamber, and the team was able to take a 3D picture and collect geological data. Deer Cave the second underground large volume of Mulu was also explored. The present chapter is an excerpt of a paper written after this field trip (Gilli 1993).

E. Gilli, *Big Karst Chambers*, Advances in Karst Science,
https://doi.org/10.1007/978-3-030-58732-1_5

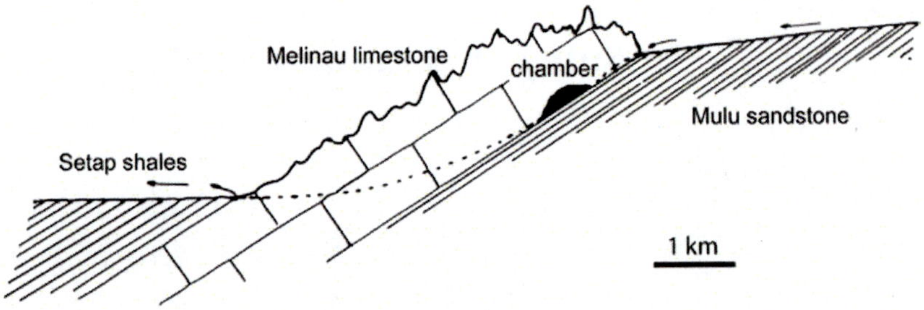

Fig. 5.1 Geological context of Sarawak Chamber

5.4 Mulu Caves

5.4.1 Location and History of Mulu Caves

Mulu karst massif is located in the NW part of Borneo in the Malaysian state of Sarawak, not far from the border with Brunei. It forms a barrier of about 30 km long, reaching an altitude of 1,600 m and extending on the west flank of Mount Mulu (2,376 m), a sandstone massif that is the highest mountain of this sector (Fig. 5.2). To the west, the limestone massif is bordered by a vast alluvial plain (altitude 65 m) (Fig. 5.3). The massif is divided into distinct compartments separated by valleys. From SW to NE, there are

- the Southern Hills,
- the Melinau Paku Valley,
- the Gunong Api (1,700 m),
- the Melinau Gorge,
- the Gunong Benarat (1,700 m),
- the Medalam Valley and
- the Gunong Buda (963 m).

The whole area is covered by a dense forest which is still inhabited by Penans a nomadic people. Mulu Caves were discovered in the 1950s by Wilford (1961) during the British Occupation of North Borneo. The true explorations took place since 1978 when several English speleological teams have succeeded on the site after the decision by the

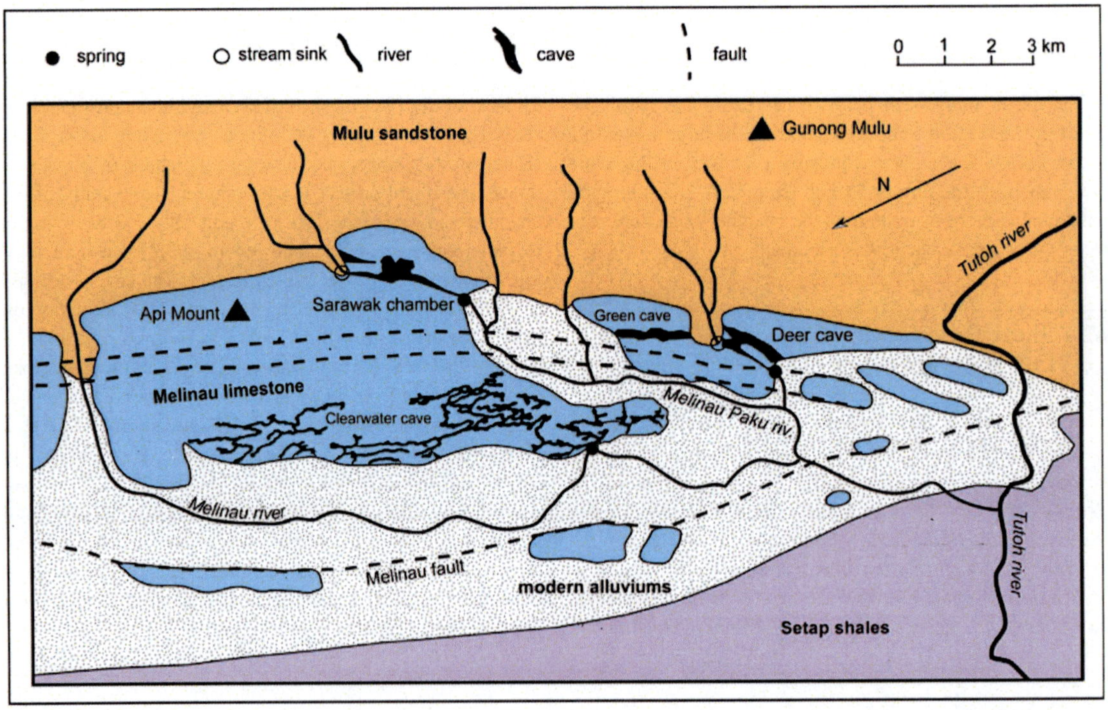

Fig. 5.2 Geological sketch of Mulu area (Webb 1981)

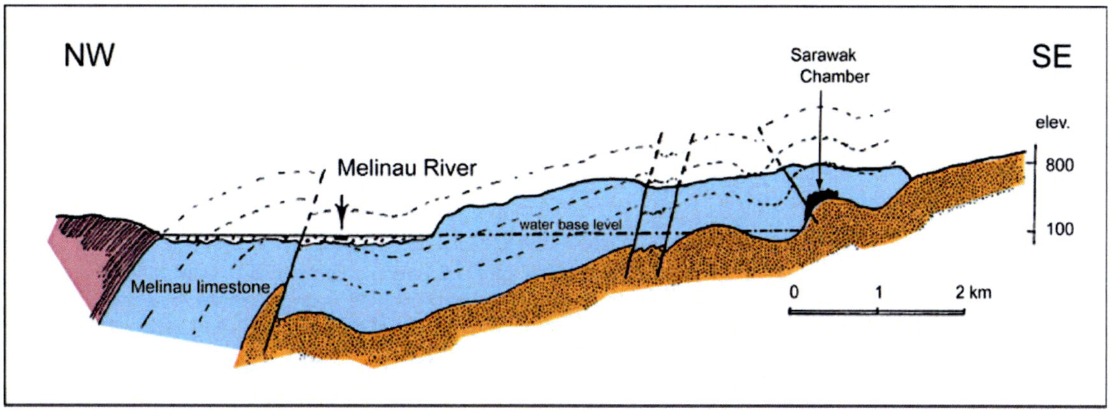

Fig. 5.3 Geological section of Mulu limestone massif

Malaysian government to create a national park. They started making an inventory of all the natural wealth of this wild jungle, including the caves. The place quickly became a high spot of world's speleology. Exploration is still in progress. The total length of explored caves overpasses 350 km. The longest cave is Clearwater cave system (227 km in 2019).

Mulu is now a national park that can be accessed by air from Miri, Limbang and Bandar Seri Begawan or by boat from Marudi. There is no road access. It is possible to visit the caves. Four caves are open to public. Other caves, including Lubang Nasib Bagus where Sarawak Chamber is, are accessible for adventure caving with a guide. Since the beginning, large volumes were encountered in Mulu massif. The largest ones are Deer Cave in the Southern Hills and Sarawak Chamber in the Mount Api.

5.4.2 Geology (After Webb 1981)

5.4.2.1 Regional Context

The limestone zone forms a line of hills at a maximum elevation of 1,700 m west of Mount Mulu. The area of Mulu comprises three main formations of Mesozoic age: the Melinau limestones sandwiched between two impervious formations, the Setap Shales and the Mulu Formation.

- The Mulu Formation (Upper Cretaceous to Eocene) is composed of clayey sandstones and shales; it takes its name from Mount Mulu where it outcrops as a vast anticline.
- The Melinau limestones (Upper Eocene to Lower Miocene) form a remarkable line of hills on the northwest flank of the previous structure. These reef limestones are bordered on the northwest by a vast alluvial plain which covers them partially.
- The whole is limited to the northwest by the hills of argillites and sandstone of Setap (Middle Miocene).

These formations were deposited in shallow areas. Their implementation is linked to the subduction phenomena between the tectonic plate of the South China Sea and that of northern Borneo. It allowed the existence of a shoal on which a barrier reef developed, around which limestones were deposited. Toward the end of the sedimentation phase, synsedimentary folds announced the beginning of a compression phase which culminated at the end of the Pliocene with the appearance of folds and faults of general direction NE/SW. In the Melinau limestones, these folds are open and associated with reverse faulting and schistosity of the same direction. After the Late Pliocene orogenesis phase, the deposits in the Mulu area were limited to superficial fluvial sedimentation, which partly overlies the Melinau limestones.

5.4.2.2 Melinau Limestone

This is a reef complex of Late Eocene to Lower Miocene age. Their thickness of about 1500 m in the Melinau Gorge decreases north and south. Briefly studied by Lietchi et al. (1960), Wilford (1961), Adams (1965), Melinau limestones were subjected to a more detailed structural analysis during speleological expeditions from 1978 to 1980 by Webb (1982) which revealed a structure more wrinkled. Lithological variations are not very marked and often masked by karstification and patina. The massive structure and the fossils confirm that it is essentially a lagoon reef. Limestone is very rich in carbonate with less than 1% impurities. The percentage of dolomite varies between 2 and 20%. The dolomitization observable in certain zones seems to be synsedimentary.

5.4.2.3 Structure

The original structure of the reef is difficult to observe. It is, nevertheless, clearly visible at the base of the series, at the Plunge Pool, in Lubang Nasib Bagus, the cave where Sarawak Chamber is located. In that place, algal formations and coral buildings are clearly visible. Stratification appears very irregular. The whole is surmounted by a mass of very compact limestones where the bedding is sometimes invisible, masked by the schistosity. Initiated as early as the Oligocene, an orogenic phase gave rise to a family of open folds and reverse faults (Fig. 5.3). The main one, the Melinau fault, cuts the massif into two distinct units; it is emphasized in the Melinau Plain by a line of limestone hills. The style of the folds is most often well opened with widely rounded anticlinal hinges and more pinched synclines.

A slight metamorphism transforming limestones locally in marble and a well-marked schistosity are linked to these folds. The post-orogenic decompression opened the schistosity planes, individualizing in the massive limestones elongate microliths, separated by parallel and tight joints with an orientation parallel to the axis of the folds. This family of discontinuities plays a considerable role in speleogenesis.

Two recrystallization zones parallel to the folds cross the massif in its length. It seems to be a phenomenon related to schistosity. At this level, the discontinuities are poorly developed. The recrystallization zones appear to form important hydrological barriers; however, given their association with the structure, it seems more logical to attribute this barrier effect to the anticlines.

5.4.3 Geomorphology

The evolution of the karst is linked to the lowering of the local hydrologic base level in the Setap Shales. They form a soft and impermeable layer that initially covered the limestones. The evolution of the massif was made as the limestone was exposed due to the erosion of Setap Shales from east to west. The different cave networks follow this trend. The groundwater deepened and shifted toward the west. It is thus very clear that the surface forms and the underground morphologies are older and more developed in the eastern part of the massif. In that area, tower karst and huge underground volumes are present.

5.5 Sarawak Chamber

5.5.1 History and Survey

Sarawak Chamber was discovered and surveyed in 1980 by Andy Eavis, Dave Checkley and Tony White during the Mulu 1980 expedition. Thinking first of all they were walking in a gigantic circular gallery, the explorers, limited by the power of their lighting, became aware of the importance of their discovery only, thanks to their survey: 600 m long, 415 m wide, an area of 16 ha for an estimated volume of 12 millions m^3.

The chamber was surveyed again with a 3D Laser Scanner, in February 2011, by Kevin Dixon during an Anglo-Malaysian expedition which confirmed more or less the previous survey (Dixon 2011). The dimensions measured in 2011 are 435 m wide, 600 m long. The ceiling is between 105 and 115 m above the floor. A circle of 325 m diameter can be fitted into the floor plan of the chamber. The data confirm that Sarawak Chamber, with its volume of 9,579,205 m^3 and area of 164,459 m^2, is currently the world's largest natural underground chamber (Figs. 5.2, 5.3 and 5.4).

5.5.2 Description

Lubang Nasib Bagus Cave includes a main gallery traversed by an underground river. This is the resurgence and the underground course of the Hulu Air Jernih River which sinks into the limestone in the Hidden Valley, 3 km north. The underground route is almost rectilinear. At 1.5 km from the entrance, the chamber extends on the right bank. The river gallery goes further, until a sump not far from the Hidden Valley. In this valley, located east of the previous swallow hole, Prediction Cave belongs indisputably to the same cave network of which it forms a dry level (Fig. 5.5).

Access to Lubang Nasib Bagus is via a 4-h walk through the jungle of Melinau Paku Valley toward the imposing cliff which forms the right bank. The high and narrow entrance is traversed by an airflow which betrays the importance of the underground volumes. The journey to the chamber is relatively easy and fast (1.5 km) if the water is low. There are three distinct parts of equivalent length:

- an entrance lake,
- a river that can be violent during floods,
- a steep scree.

This last part enlarges quickly to reach 80 m wide at the entrance of the room. The floor of the hall is very inclined toward the west. It consists of a blockfield poorly stabilized. The ceiling is 100 m high, rather flat and roughly parallel to the floor. It is difficult to get an overall view of Sarawak Chamber, especially since the floor is very irregular and two scree ridges limit the field of view from the upper parts (Fig. 5.6).

The room can be divided into three sectors:

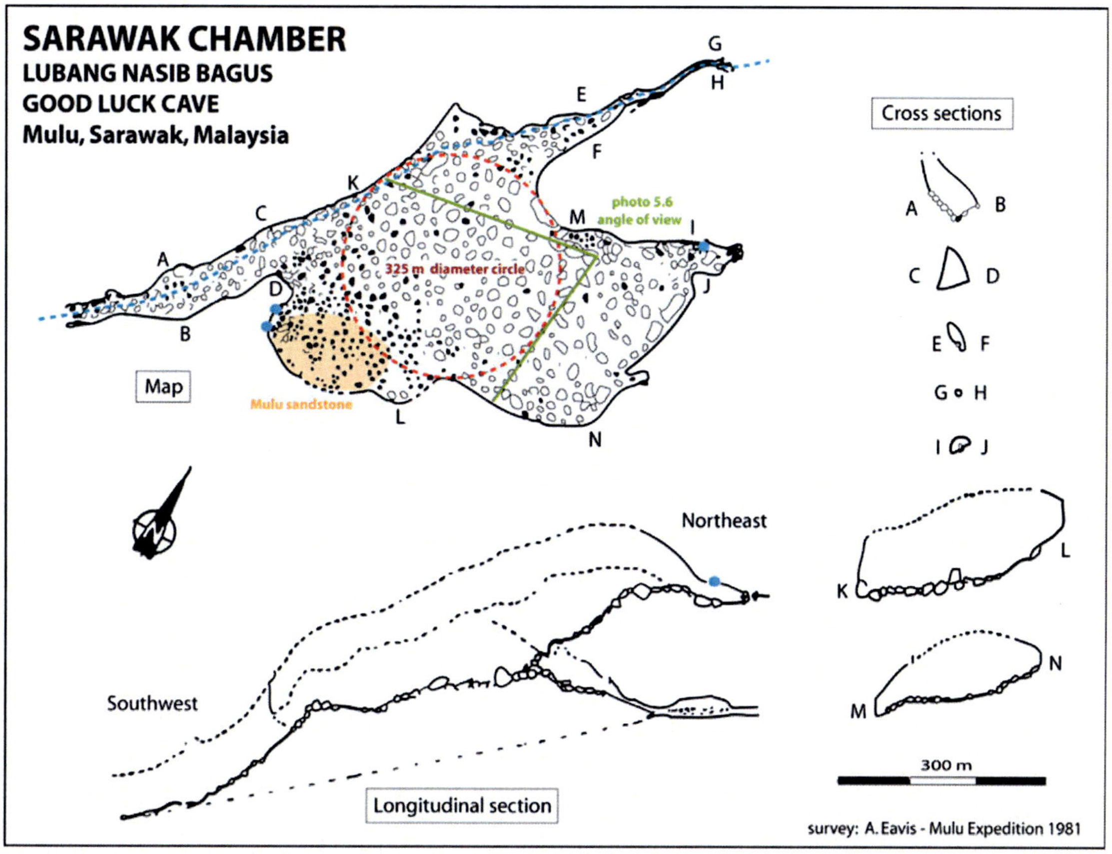

Fig. 5.4 1980s survey of Sarawak Chamber (Eavis 1981)

- to the east, an enormous blockfield which forms the bottom of the room and under which the underground river flows;
- to the south, a steeply sandstone slope more or less covered by scree;
- to the northeast, a blockfield, partly covered with flowstone.

A visit to the chamber starts by climbing a slope of unstable limestone scree, very upright, to a stream that pours from a joint in the south wall. In this sector, outcrops of Mulu Sandstone are visible on the floor. From this point, 50 m to the northeast on a flat, a crater 10 m in diameter was created into the sandstone by the fall of a huge block from the vault (Fig. 5.7). Unlike the rest of the cavity, the total absence of guano on this block shows that it was a very recent fall when discovered in 1993. This means that the chamber enlargement is still in progress.

Crossing the flat area to the northeast leads to a new steep slope until the north wall. The path is between several meters large blocks; then the slope decreases, and the blocks are smaller and more or less covered with flowstone. To the north, a gallery is located in the extension of Prediction Cave, but no junction could be achieved although a strong airflow confirms a link.

5.5.3 Structure of Sarawak Chamber

The walls are formed by discontinuities planes attributed essentially to the schistosity. Karst deposits and darkness complicate the observation, and the dip is difficult to apprehend. The general structure of the chamber is complex; however, the examination of the 3D pictures (Fig. 5.6) taken during the short mission allowed the cavers to approach the large lines (Figs. 5.8 and 5.9).

The southern part shows Mulu Sandstone outcrops on the floor. It is likely that these sandstones constitute the entire floor of the room, partly covered by the limestone blocks collapsed from the vault. The presence of these outcrops and

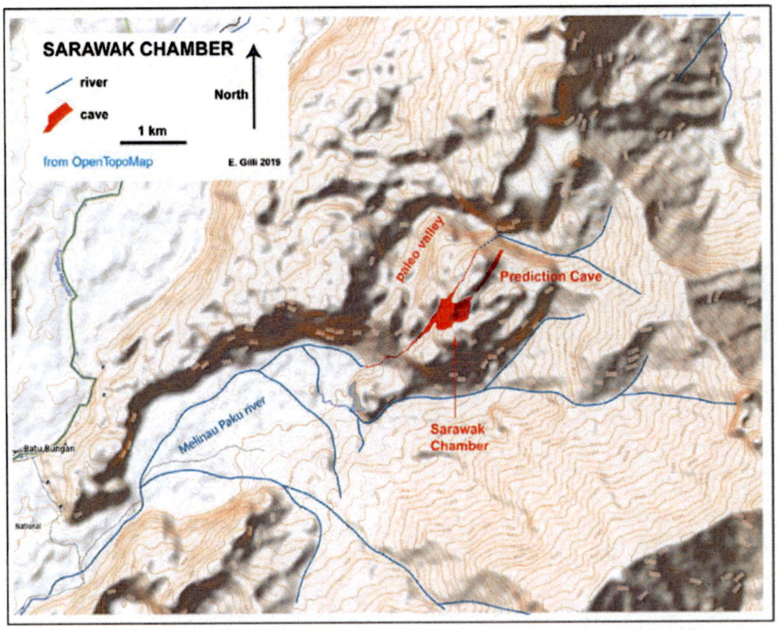

Fig. 5.5 Topographic map of Sarawak Chamber area

Fig. 5.6 General view of Sarawak Chamber. The picture was realized in 1993 using magnesium powder. The arrows show two cavers and a tent, the angle of view is in Fig. 5.4

Fig. 5.7 Recently collapsed rock in Sarawak Chamber

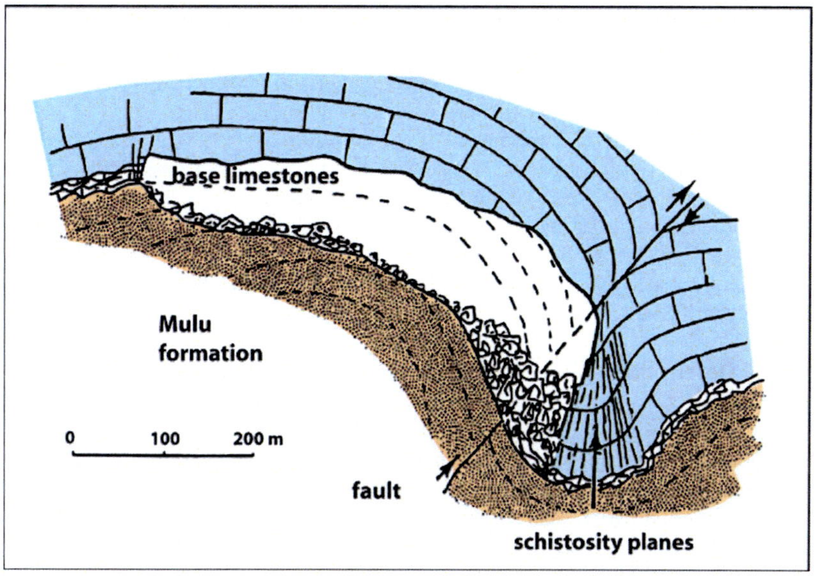

Fig. 5.8 Structure of Sarawak Chamber

the examination of the stereo photos indicate that the room is dug in an anticline, which confirms the surface observations of the English geologists in the 1980s (Webb 1982).

5.5.4 Chamber Genesis

The cave was dug by the Hulu Air Jernih River. Its watershed extends on the impervious terrain of the northwestern side of Mulu Mount. A huge amount of water is collected because rainfall may reach 10 m per year. The river met the Melinau limestone at the Hidden Valley and sunk to the SW through an anticline. It started digging Prediction Cave, the beginning of Sarawak chamber and Nasib Bagus downstream gallery. The pictures of the room clearly show a gallery above the main gallery that could be a witness of this digging stage. Then the underground river flowed to the west to the parallel syncline and emerged in the Melinau Paku Valley. As the river had reached the substratum, it is probable that the enlargement of the cave by scouring of the sandstones had begun.

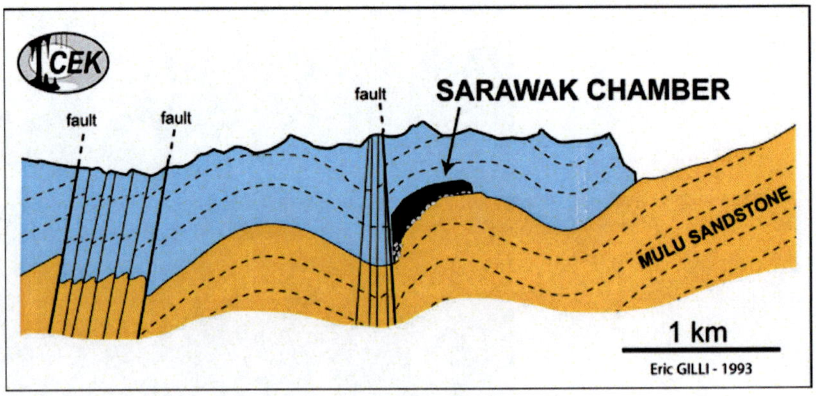

Fig. 5.9 Structure of Api Mount

Then, the Melinau Paku deepened its bed, which increased the hydraulic potential; the karst water circulations sank quickly with a shift to the west running on the steep flank of the anticline. This favored the excavation of a pre-chamber within the sandstone, the deepening of Hidden Valley and the creation of the current sink. Prediction cave became a dry gallery. The pre-chamber was no more crossed by the river; however, it was still excavated in the west by lateral scouring of the sandstone by the underground river. A fast withdrawal of coarse elements is possible downstream, thanks to the river speed. Indeed, it is almost rectilinear with an average slope of 7%. Nowadays, in the chamber only a small stream remains, the role of which cannot be totally neglected (Fig. 5.10).

5.6 Deer Cave

5.6.1 Description

Deer Cave is one of the world's largest galleries. Easily accessible, it is the main underground tourist attraction of Mulu National Park. The cave is populated by a large colony of bats estimated between 1 and 3 million, which constitutes an additional curiosity. A path along the western edge of the limestone massif leads first to a resurgence where the water runs in a small gallery, and then about 200 m further to the monumental entrance of Deer Cave. The floor is covered with a huge blockfield crossed by a passage to the north. Then the path goes along a small stream and gets to a huge flat gallery, that is, 160 m wide. Waterfalls falling from the top of the ceiling, 120 m above the floor level, make this show unforgettable (Fig. 5.11). At the end of this huge gallery, a narrower passage (only 50 m wide!) makes it possible to reach the northern part of the cave which runs the stream that has hollowed the cave. The total length is 2.16 km (Fig. 5.12).

5.6.2 Structure

Aerial pictures show the folded structure. They suggest that the main gallery of Deer Cave extends below an anticline (Fig. 5.13). The internal structure of the cave is governed by the vertical schistosity planes that are the main discontinuities. However, it seems obvious that the stability of the ceiling benefits from this natural arch structure, like for Sarawak Chamber.

5.6.3 Genesis

The northern entrance of Deer Cave opens into the "Garden of Eden", a large oval sinkhole 1 × 1.5 km in dimension. On the other side of the sinkhole opens Green Cave, a 3.4-km-long dry cave (Fig. 5.14). Deer Cave and Green Cave are two parts of the same cave network currently separated by the sinkhole. The current stream observed in Deer Cave is a tributary of the Melinau Paku River that remained active after the migration of the main river toward west. This tributary seems too small to explain the digging of these gigantic volumes. Both caves were probably excavated by the Melinau Paku or another tributary before it moves to the west (Osmaston and Sweeting 1982). The Garden of Eden could then be interpreted as a collapsed room comparable to Sarawak Chamber (Fig. 5.15).

5.7 Conclusion on Mulu Caves

In Mulu, both underground volumes are stable, thanks to the intrinsic qualities of the reef limestones slightly metamorphosed during the folding phases and devoid of clayey interbeds. In addition, they are in a favorable structural context where anticlines provide natural vaults. At this level, the high lateral stresses related to folding favor the blocking

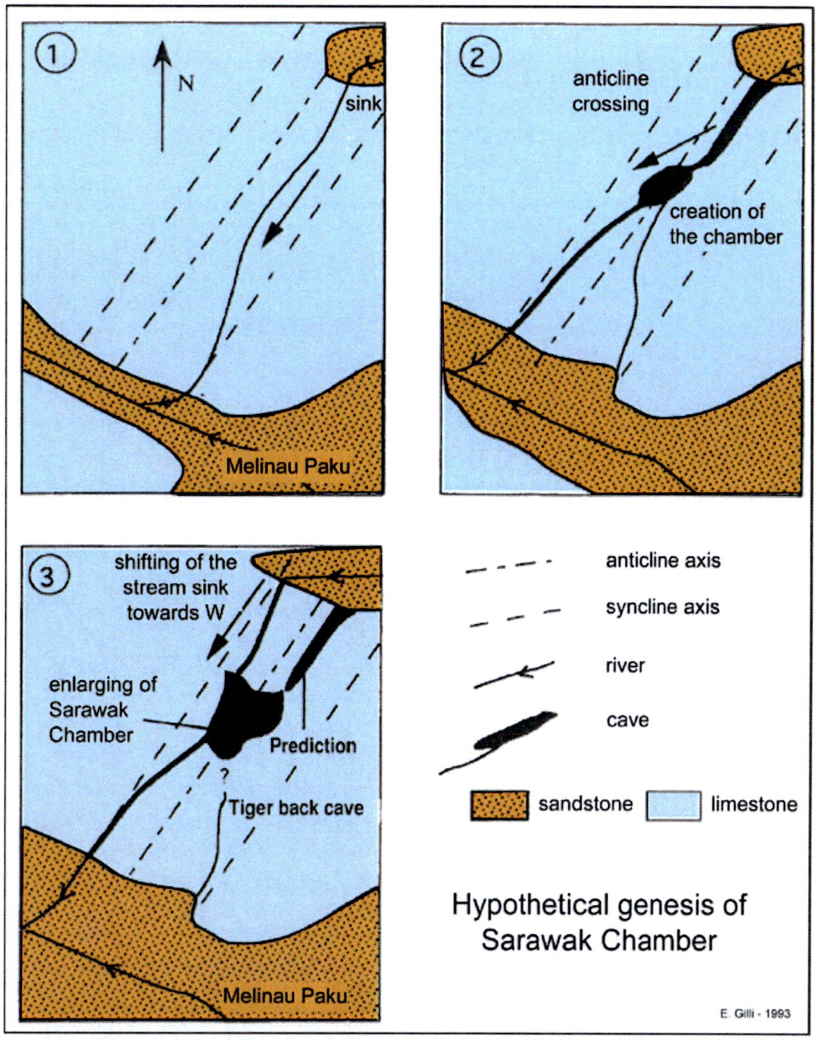

Fig. 5.10 Hypothetical genesis of Sarawak Chamber

Fig. 5.11 Deer Cave. The white lines are waterfalls

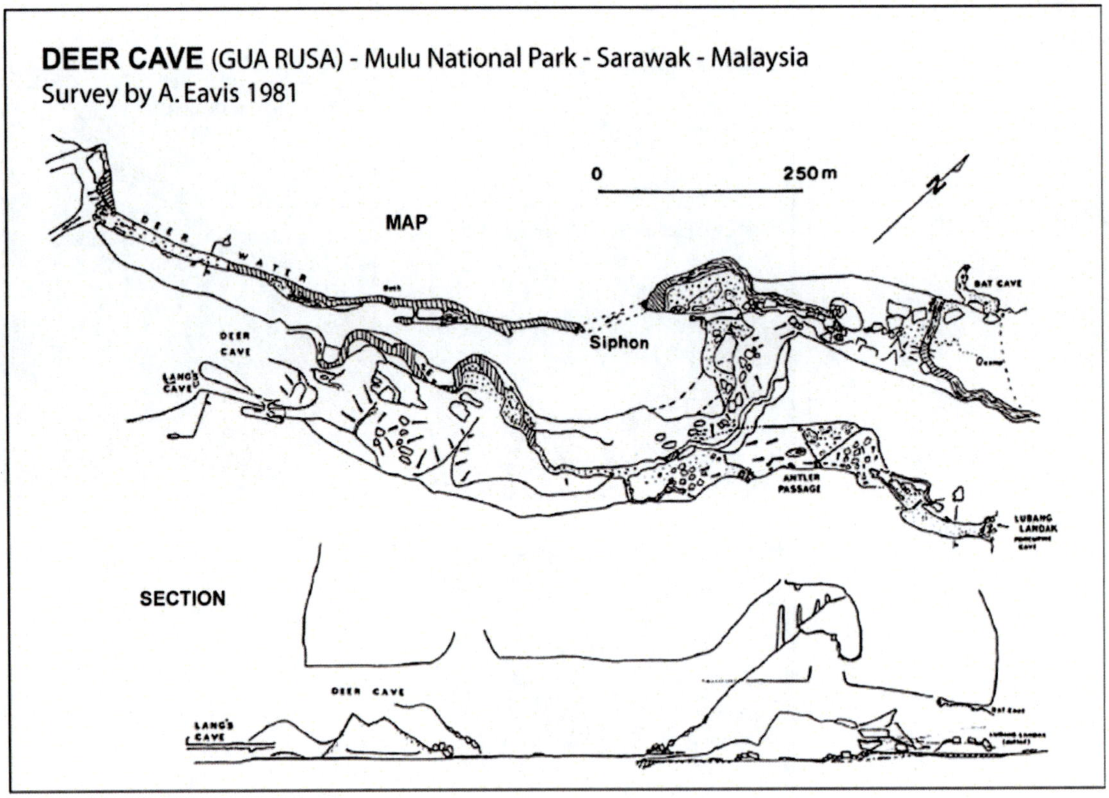

Fig. 5.12 Deer Cave (Eavis 1981)

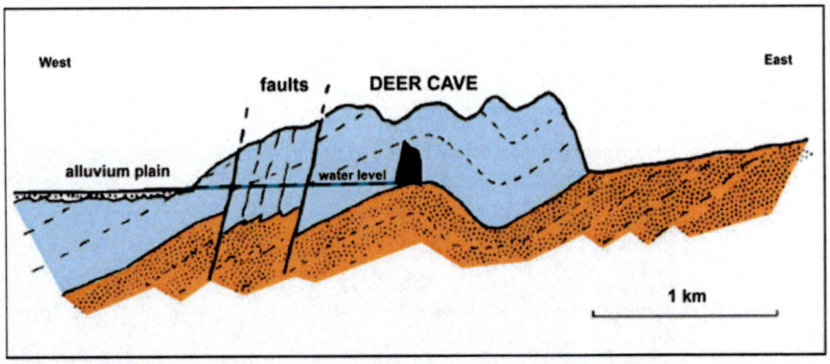

Fig. 5.13 Geological sketch of Deer Cave area

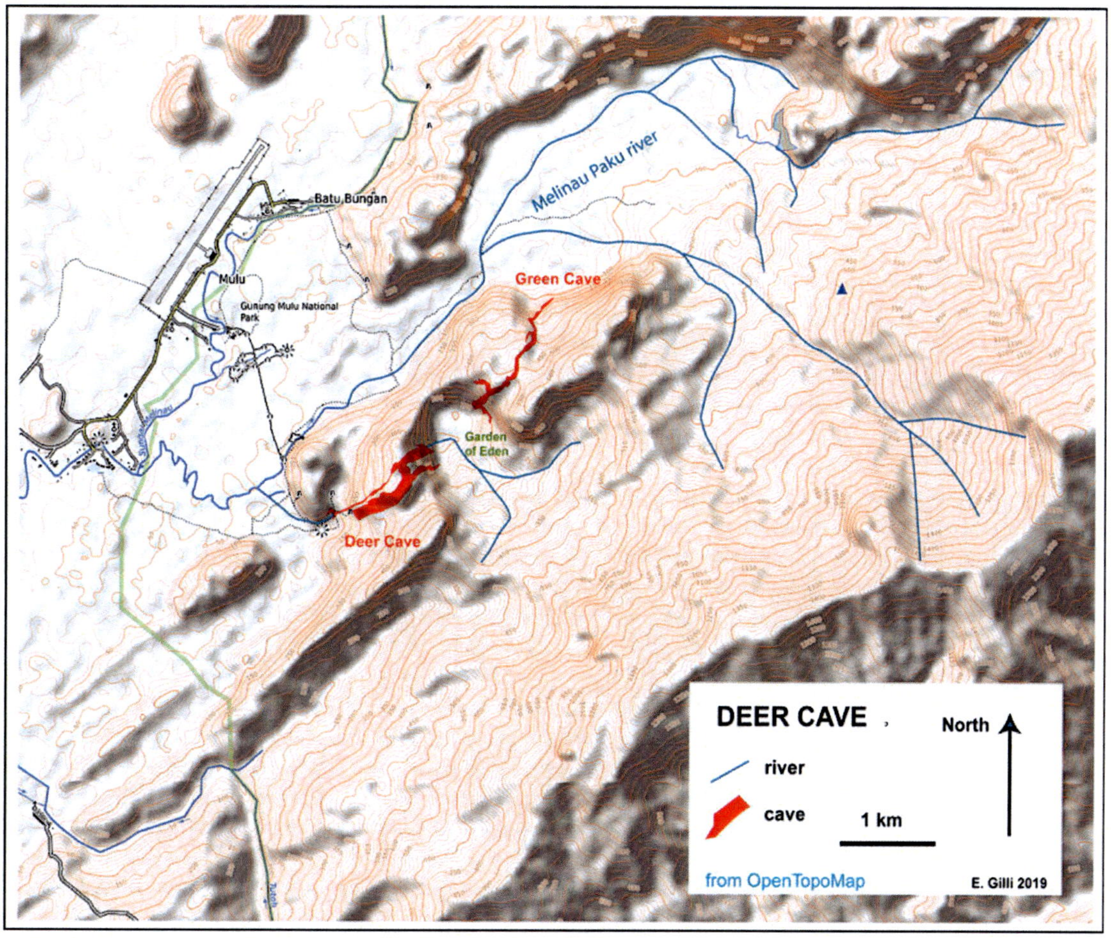

Fig. 5.14 Location map of Deer Cave and Garden of Eden

of the elements of the ceiling. This is all the better as the main discontinuities are vertical, related to the schistosity and not horizontal related to the stratification. This verticality determines the existence of elongated microliths that prevents any possibility of rotation. A rupture would then result from a destruction of the element by compression; however, it is probable that the metamorphism of these limestones confers to these vault elements an important resistance to compression allowing the considerable spans observed. Sarawak Chamber obeys these exceptional rules, and its dimensions can be explained by the presence of an anticlinal dome whose heart made of sandstone was gradually hollowed out by the river that runs through the cave. The basal layers of the limestone series, with irregular and

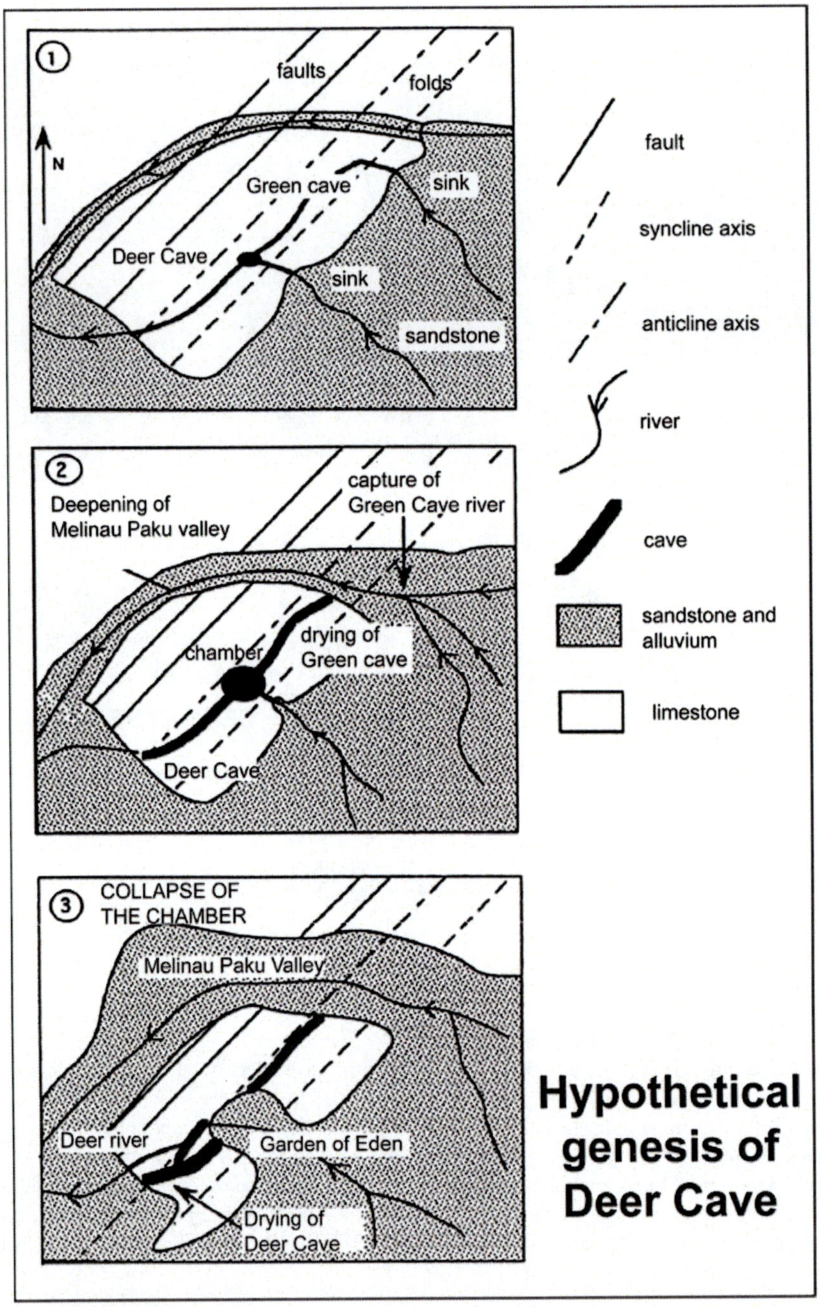

Fig. 5.15 Hypothetical genesis of Deer Cave

sometimes intersecting stratification, collapsed. The stability of the vault was acquired, thanks to the upper massive limestone beds. Meanwhile, the collapses are still active and during the cavers' 5-day camp inside the room, several rockfalls were heard.

References

Adams, C.G.: The foraminifera and stratigraphy of the Melinau limestone, Sarawak, and its importance in Tertiary correlation. Quart. J. Geol. Soc. Lond. **121**, 283–338 (1965)

Dixon, K.: Measuring Sarawak Chamber. The Mulu caves project: exploring the world's largest cave (2011). www.mulucaves.org. Accessed 9 July 2019

Eavis, A.: Caves of Mulu 80. In: Mulu' 80, Royal Geographical Society Report (1981)

Gilli, E.: Les grands volumes souterrains de Mulu (Bornéo, Sarawak, Malaisie). Karstologia **22**, 1–14 (1993). https://doi.org/10.3406/karst.1993.2312

Lietchi, P., Roe, F.W., Haile, N.S.: The geology of Sarawak, Brunei and the western part of North Borneo. Bull. Geol. Surv. Dept. Terr. Borneo **3**, 2 (1960)

Osmaston, H.A., Sweeting M.: Gunong Mulu National Park: geomorphology. Sarawak Museum J. **XXX-51**, 75–93 (1982)

Webb, B.: Geology. In: Mulu' 80 Expedition Royal Geographical Society Report (1981)

Webb, B.: The geology of the Melinau limestone of the Gunong Mulu National Park. Cave Sci. **9–2**, 94–99 (1982)

Wilford, G.E.: The geology and mineral resources of Brunei and adjacent parts of Sarawak. Mem. Geol. Surv. Dept. Terr. Borneo **10**, 319 (1961)

6 The Future of Underground Plants

This chapter presents examples of artificial chambers large enough to host a big plant. They support the feasibility of underground nuclear power stations.

6.1 Current Examples of Large Underground Volumes

6.1.1 Underground Plants

Natural examples prove that it is possible to have caverns large enough to host underground plants, and man-made examples also show their feasibility at a lower scale than the natural examples. The first cavern for a hydropower plant was excavated in 1898 at Snoqualmie (Washington, USA) (Fig. 6.1). Nowadays, hundreds of plants exist; however, for most of them the span of the ceiling is less than 20 m (Fig. 6.2). The French Sautet Cavern was for a long time an exception with a 36 m span (Fig. 6.3).

There is only one example of an underground nuclear plant (UNPP). The Lucens plant in Switzerland was an experimental station installed in a circular cavern 25 m high and 20 m in diameter (Fig. 6.2). It was destroyed in 1969 after a problem on the cooling system that provoked the meltdown of the reactor core. The underground construction prevented a massive diffusion of the radioactivity outside (IFSN 2014) which can be compared with the effects of Chernobyl or Fukushima accidents.

Since the 1950s, the French engineer P. Duffaut took his pilgrim staff to share his convictions during meetings and international conferences trying to promote UNPP (Duffaut 1954, 1995, 2017). He inventoried several important underground realizations. He also asked the International Society for Rock Mechanics (ISRM) in 2013 to create a dedicated commission on the underground siting of nuclear reactors. The commission delivered the first report in 2015 saying that the technical feasibility of UNPP is real. However, as far as is known, no project is currently in progress.

6.1.2 Underground Cavern Hall for Public Use

In Norway, in the late 1970s, a study focused on the possibility of digging a chamber wide enough to host a standard nuclear reactor. An experimental site was chosen in Liåsen, but Norway banned the use of nuclear energy which stopped the project. It was renewed in time for the 1994 Winter Olympics, and a 61-m-span cavern was excavated in Gjøvik, for an underground skate ring (Fig. 6.4) (Duffaut and Sakuraib 2018; NFF 2017).

6.1.3 Underground Quarry

The world's largest example is Tytyri where excavation caverns have a span larger than 100 m. The structural environment with favorable horizontal constraints was used by local engineers to adapt the shape of the cavern during the digging (Figs. 6.5 and 6.6).

6.2 General Conclusion

The objective of the thesis work on karst underground volumes was to study the possibility of digging artificial volumes large enough to shelter a nuclear power station. This work was complemented with a further study on Sarawak chamber in Borneo. Both works showed that very large volumes can be stable in specific geological environments.

Since this naturalistic research no other work on that topic was done, man-made examples show that the hollowing of very large underground caverns is now possible. Jules Verne, the famous French science fiction author, described underground worlds in *Voyage au Centre de la Terre* (Journey to the Center of the Earth—1864) or an underground city in *Les Indes Noires* (The Black Indies—1887). His utopia turns in realty.

Very large caverns could welcome large plants, for instance, nuclear ones. An underground solution is favorable

E. Gilli, *Big Karst Chambers*, Advances in Karst Science,
https://doi.org/10.1007/978-3-030-58732-1_6

Fig. 6.1 Snoqualmie Cavern. The first underground power plant in the world (Bowen 1900)

to protect the plants against vandalism, war or terrorism. Furthermore, a subterranean location is also favorable if an earthquake occurs. Indeed, the destructive effects of an earthquake augment when the seismic waves reach the contact between the rock and the atmosphere because the buildings can oscillate dangerously independently of the ground. In

UNDERGROUND POWER PLANTS

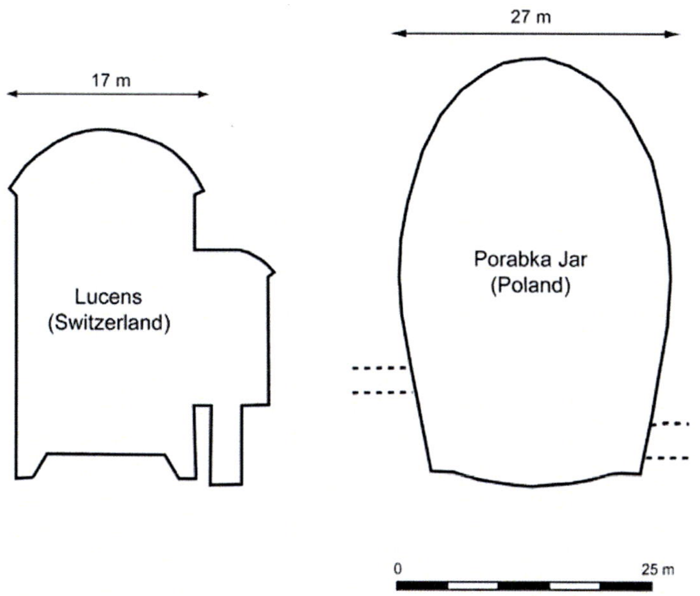

Fig. 6.2 Lucens Cavern (Switzerland) and other ones in Poland and Malaysia (Duffaut 2018)

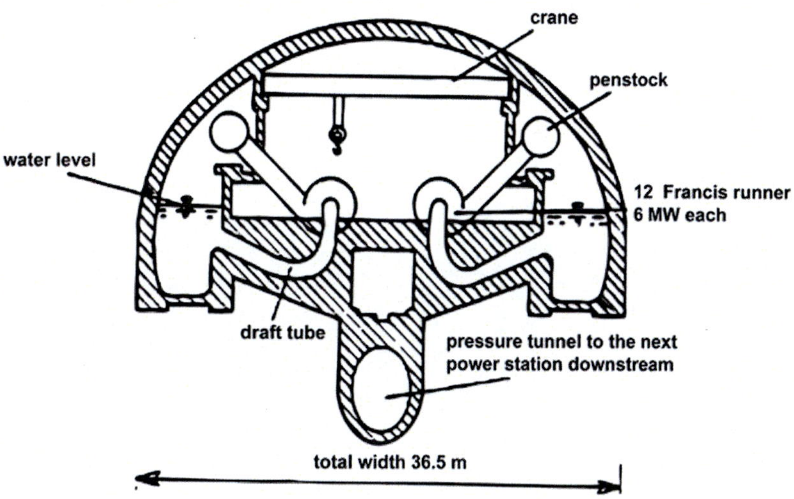

Fig. 6.3 The Sautet underground powerplant (France). World's widest example in limestone

contrast, deep underground structures are supposed to oscillate with the surrounding rock. Observations on tunnels confirm that underground structures are less vulnerable to earthquakes than superficial ones (Hashash et al. 2001). However, this depends probably on the size and the elasticity of the underground equipment. Whatever it is, underground plants would be more protected against earthquakes and tsunamis than outside ones.

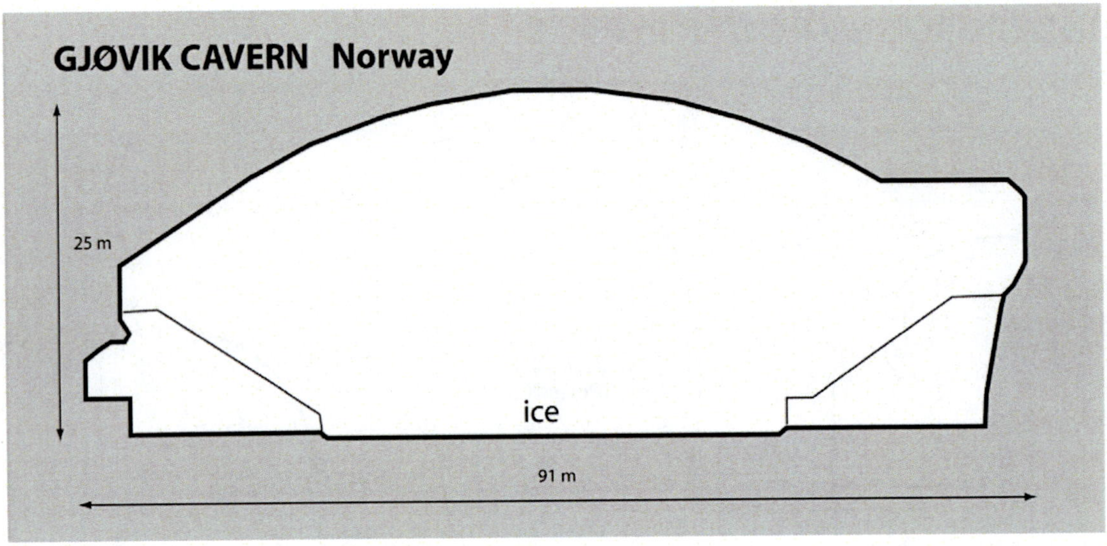

Fig. 6.4 Underground ice ring in Gjovik Cavern (Norway)

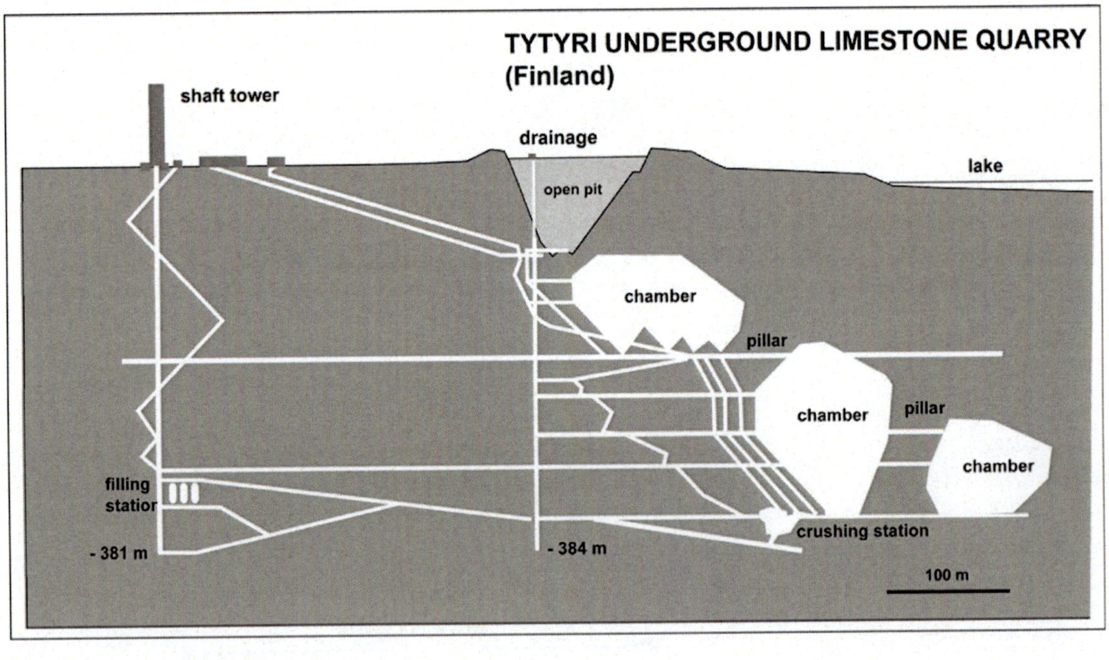

Fig. 6.5 Cross section of Tytyri underground quarry

Fig. 6.6 View of one chamber in Tytyri underground limestone quarry

References

Bowen, A.D.: Seattle and the Orient. The Times Printing Company, Seattle (1900)

Duffaut P.: Réflexions de géologie appliquée sur les excavations souterraines de grandes dimensions, rapport interne EDF Service Géologie, p. 17 (1954)

Duffaut, P.: Compte-rendu du Symposium de Gjovik. Tunnels et Ouvrages souterrains **125**, 141–145 (1995)

Duffaut, P.: Les réacteurs souterrains: une bonne idée trop vite enterrée? RGN, pp. 48–51 (2017)

Duffaut, P., Sakuraib, S.: Back to underground reactors. In: 10th Asian Rrock Mechanics Symposium, Singapore, 29 October to 03 November 2018

Hashash, Y.M.A., Hook, J.J., Schmidt, B., John, I., Yao, C.: Seismic design and analysis of underground structures. Tunn. Undergr. Space Technol. **16**, 247–293 (2001)

IFSN (2014): Centrale nucléaire expérimentale de Lucens. https://www.ensi.ch/fr/themes/centrale-nucleaire-lucens/ Accessed 28 Jan 2020

NFF (2017): The principles of Norwegian tunnelling. Norwegian Tunnelling Society, publication no 26, pp. 84